P9-DXI-562

Safely to Earth

UNIVERSITY PRESS OF FLORIDA

Florida A&M University, Tallahassee
Florida Atlantic University, Boca Raton
Florida Gulf Coast University, Ft. Myers
Florida International University, Miami
Florida State University, Tallahassee
New College of Florida, Sarasota
University of Central Florida, Orlando
University of Florida, Gainesville
University of North Florida, Jacksonville
University of South Florida, Tampa
University of West Florida, Pensacola

SAFELY

University Press of Florida
Gainesville · Tallahassee · Tampa · Boca Raton
Pensacola · Orlando · Miami · Jacksonville · Ft. Myers · Sarasota

TO EARTH

THE MEN AND WOMEN WHO BROUGHT THE ASTRONAUTS HOME

JACK CLEMONS

Frontis images: Apollo 17 splashdown and Space Shuttle *Endeavor* landing.
NASA images.

Printed in the United States of America on acid-free paper

This book may be available in an electronic edition.

23 22 21 20 19 18 6 5 4 3 2 1

Library of Congress Control Number: 2017964061
ISBN 978-0-8130-5602-9

The University Press of Florida is the scholarly publishing agency for the State University System of Florida, comprising Florida A&M University, Florida Atlantic University, Florida Gulf Coast University, Florida International University, Florida State University, New College of Florida, University of Central Florida, University of Florida, University of North Florida, University of South Florida, and University of West Florida.

University Press of Florida
15 Northwest 15th Street
Gainesville, FL 32611-2079
http://upress.ufl.edu

To Denise, for her love and support during our many adventures together, and for always returning me safely to Earth

In a very real sense, it will not be one man going to the moon . . .
it will be an entire nation. For all of us must work to put him there.

President John F. Kennedy to a joint session of Congress
May 25, 1961

CONTENTS

Safely to Earth

1

PRE-FLIGHT BRIEFING

On November 14, 1969, lightning struck Apollo 12's Saturn V rocket at liftoff at the Kennedy Space Center, knocking many of the spacecraft's critical electrical power systems offline. This set off emergency alarms and a flood of warning lights, and caused a vast amount of erroneous and inexplicable data to be displayed to both the astronauts and the ground controllers in Houston. It also left the Apollo Command Module, now departing for the Moon, with batteries capable of only about three hours of power.

Neither the astronauts nor the Mission Control team knew about the lightning strike, so they were in the dark about what had happened. Only quick and clear thinking by NASA's John Aaron, a young flight controller in Houston's Manned Spacecraft Center, brought the onboard systems back online and avoided a potentially catastrophic mission abort.

John had recognized a pattern in the inconsistent numbers being displayed that indicated one of the onboard electronics systems had shut down. He quickly requested that the crew place an obscure cockpit switch, called Signal Conditioning Electronics, into its Auxiliary position. Once the SCE was switched to AUX, all of the spacecraft's correct data were restored, revealing the underlying problem, which allowed John to inform the crew of the actions needed to restore full power to their spacecraft. Apollo 12 went on to carry the third and fourth human beings to a walk on the Moon and return them safely to Earth.

John Aaron is just one of the remarkable professionals who were central to achieving this nation's voyage of humans into space, landing on the Moon, and later, making the Space Shuttle fly. He and many other outstanding men and women played critical roles in the early success of NASA's manned space program flights, and more important, ensured that the astronauts who flew them returned home safely. The stories that follow will acquaint you with a number of these individuals, and reintroduce you to John. Many such extraordinary people inhabit this account, a diary that chronicles my time working beside them as part of that adventure.

2

PREP FOR LAUNCH

"But why, some say, the Moon?"

On September 12, 1962, President John F. Kennedy gave his now famous "We choose to go to the Moon" speech at Rice University in Houston, Texas. On that sweltering day he asked his audience a question: "But why, some say, the Moon?" I've been asked a version of this question many times over the years. I've heard enthusiasts and apologists alike respond by citing the thrill of exploration, the Cold War, a quest for scientific understanding, the commercialization of Teflon, and the invention of integrated circuits. (Neither of the latter two is right, though one study found that every dollar spent by NASA on research and development ultimately returns seven dollars of value to the gross national product.)

But that day President Kennedy answered his question with what for me is the most compelling argument: "Because that goal will serve to organize and measure the best of our energies and skills." So it did. By accepting that challenge, NASA and the nation energized the imagination and spirit of a younger generation of Americans who embraced the study of engineering and the sciences, and many of whom took part in the president's grand enterprise. I was one of those. So when I'm asked what benefit came from the Apollo Program, I now respond, "Me and hundreds of others like me, who answered the President's call."

Apollo was the great adventure of our age. Though numerous books and films have celebrated the achievement, when people discover I had a part in it, their first question is often "What was it like?" Even after all this time, interest in Project Apollo seems undiminished. My answer is a personal one. I describe how I experienced the program as a newly minted "rocket scientist" and offer a glimpse through the narrow window my role provided into an unprecedented human undertaking. I find that when I talk about the tasks assigned to me on Apollo, and the world of technology available in the 1960s, people get a sense of the rest of it. And an appreciation for the staggering responsibility shouldered by the men and women of NASA and the private contractors who brought those parts together and put human footprints on the Moon.

Because, of course, I was only one of many. The efforts of the teams I worked with were integrated with those of scores of professionals, all applying their specialized skills to the construction of these systems. Hundreds of thousands of men and women, working in offices and factories and research centers and launch sites and tracking stations and military bases and on Navy ships around the world, added their talents to make Apollo a success. They were largely unfamiliar with our work; they had to trust us to take the same care as they did to ensure that all components worked together when called upon. When at last human footprints were imprinted on the Moon, many of those people who had pioneered mankind's first ventures into space went on to apply their hard-won experiences to developing the Space Shuttle, a way for humans to work and live in space and study the wondrous universe from a vantage point once only dreamed of.

I'm proud of the programs on which I worked, of the people whom I met and learned from and became friends with, and of what we accomplished for human spaceflight. I'll describe some of that work here; sketch a few portraits of those unknown pioneers; put on exhibit the tools they used—many now considered primitive—and explain the risks they took and the mistakes they made in producing these brilliant creations. These are little told tales of the people who fashioned the computer programs that accompanied the human crews on these spacecraft, and of the astronauts who relied on them for the directions to their destination, for the guidance on how to get there, and for control of their fate during the journey—and, when the space travelers completed their often perilous tasks, to return them safely to the Earth.

In the opening of his 1947 novel *Tales of the South Pacific* James Michener wrote:

> They will live a long time, these men of the South Pacific. They had an American quality. They, like their victories, will be remembered as long as our generation lives. After that, like the men of the Confederacy, they will become strangers. Longer and longer shadows will obscure them until their Guadalcanal sounds distant on the ear, like Shiloh and Valley Forge.

So, I fear, could be the fate of the men and women who committed their energies and passions to provide humankind the promise of journeying beyond the narrows of Earth. As the years pass into decades and the technology we used seems increasingly akin to Stone Age hammerstones and flint flakes, let these people not become strangers.

3

COUNTDOWN

I'm a Dreamer, Not an Engineer

"For my sixth birthday, my dad gave me a brand new Erector Set, and from that day on I knew I wanted to be an engineer!" Nope. That was never me.

A popular misconception is that we rocket scientists are a nerdy bunch, obsessed from the time we learn to read with building toys that fly. While that was the case for some, including an astronaut friend of mine, it wasn't true for me or for a number of other engineers and technical people I know. It wasn't our degree, but rather an early enthusiasm for human spaceflight that we shared, and which, given the opportunity, propelled us to be part of it. As an example, here's a little background to illustrate the improbable route I took to work on the Apollo Moon Program. I know a number of professionals who traveled equally unlikely paths.

It is true that I got an Erector Set when I turned six. I remember unwrapping the rectangular metal box, painted red except for a drawing on the lid of a kid my age staring at some metal contraption he'd put together. I unlatched the hasps and lifted the lid and found a mystifying assembly of thin metal beams with holes drilled through them, and wheels, and a whole lot of screws, and a key-wound motor, and some printed directions, all neatly tucked into the spaces inside. I played with it for a time; I built a few simple towers that wouldn't quite stay upright, and a wind-up crane that didn't lift. I wasn't good at getting the joints to fit so I spent more time constructing wobbly metal toys than playing with them. After a couple of weeks I put everything back in the box and never opened it again. I don't remember seeing it after third grade when we moved to the other side of town.

I didn't build model rockets when I was a teen either, though one of my friends did and I watched him fire one off, worrying the wind would blow it backward and onto me. I did have a chemistry set that I kept in the basement, but I only remember a concoction of powders that I mixed together in a mason jar. Somehow it got wet and exploded one day while I was at school; my mom threw out the box and its chemicals before I got home. I didn't miss them.

My dad did all the repairs around the house; he even converted our attic space into a really nice bedroom for my sister. I tried helping him once or twice, but both times he ended up yelling at me to just get out of the way. He used to tell me, "You better get an education, because if you have to earn a living using your hands you'll starve to death." He said that quite a lot.

So, no, I didn't always want to be an engineer—in fact, the thought of actually building something was daunting. Mostly I was a reader. I started with comic books, usually Batman and Superman, but some Scrooge McDuck too. My dad had been a state policeman for a time and he opened his own private detective agency when I was in middle school. That's when I discovered the Hardy Boys—I read all thirty-four books that were available when I was eleven; I guess I was hoping my dad would take me out with him on one of his cases, like Fenton Hardy did with his sons. My father wasn't persuaded, and that didn't happen either.

I even wrote ten (mercifully) short stories called The Adventure Boys, complete with my hand-drawn illustrations that featured me and my best friend at the time, Jim Zeigler, solving crimes with my father. One story had the melodramatic and derivative title "The Clue of the Feathered Signet." I still have a couple of those, which I guess is a measure of how, even then, my interest was in writing and dreaming, not in building.

By the time I reached middle school I'd outgrown both comics and the Hardy Boys, and a classmate friend introduced me to science fiction. I discovered Isaac Asimov's robots, Ray Bradbury's Mars, and Arthur C. Clark's Moon. I was transfixed. It wasn't the science so much as the call to adventure that appealed to me, though the likelihood of it ever being "real" back then wasn't too removed from believing that Batman could fight crime from a cave.

In 1957 my older sister Mary married a young man named Bob Sunday, who literally changed my views on that. Bob was a U.S. Army veteran who had served in combat duty as a tank driver in the Korean War. He was a professional artist, a skilled concert organist, an avid photographer, and most important for me, an enthusiastic amateur astronomer. Bob was twelve years my senior, and when I was an early teen, he taught me to use his telescope and showed me wonders of the night sky—the stars of the Pleiades, the rings of Saturn, and the craters on the Moon. I've credited him often as being one of the two most influential people in my life, the other being John F. Kennedy, who inspired my interest in space and fueled my desire to be part of the team who put human beings on the Moon.

That's where things stood when I went off to college in 1961. After my sister married, my mother and father moved from Pennsylvania to Florida in 1958 so that my dad could take a police officer's job in a small beach town. I was a good enough student in high school to get a four-year scholarship to the University of Florida, but I was a thimble-fingered dreamer with little appetite for building gadgets and little skill at other pursuits, and I was still adrift about what I wanted to be when I grew up. I could do advanced math well enough when I put my mind to it, but I never liked it—an indifference that continues to this day. Yet six years later I was an engineer working to get the first human beings on the Moon safely home.

My message is, judge your future engineer by her passions, not by her toys.

Chasing Apollo

Two events occurred while I was in college that changed the direction of my life. The first was in 1962, when I was starting my sophomore year. I still wasn't sure where to direct my studies. In those days you didn't have to make that decision until you started your junior year; freshman and sophomore years were dedicated to rounding out your education: history, the humanities, writing and public speaking, chemistry, and economics all were part of the curriculum. I knew that a college degree was the key to unlock many doors, but I didn't know which ones I wanted to open. I thought about majoring in writing, though it was certainly not a profession where one might earn much of a living, unless I could become the next Bradbury or Asimov.

Then on September 12, 1962, John F. Kennedy challenged us, all of us, to go to the Moon, and I was electrified. The future I'd read about, that I'd so often imagined, was suddenly on its way and I could be part of making it happen! As I said, I'd done okay in math and science, but that was just schoolwork where I'd learned to pass the tests. Now I had a spur for my enthusiasm, a direction; the mysterious and challenging specifics of the advanced sciences and complex calculus were tools I'd have to master if I wanted to play a part in it. I changed my major to aerospace engineering (then a newly created program of study) and learned to compute the odd behavior of satellites in orbit, and to calculate the heat of friction on the surface of a supersonic aircraft. If I got my degree in time, I could be part of achieving President Kennedy's goal, "before this decade is out, of landing a man on the Moon and returning him safely to the Earth."

The second change that happened that year was that I made an unexpected, and ultimately aborted, decision to become an officer in the United States Army. Our country's engagement in the Vietnam War started in 1961, and all young men over the age of eighteen were required to register for the draft. The war escalated all through the 1960s, and so did the number of men who were drafted. When I arrived at Florida in 1961, the biggest campus rage against authority was the so-called "panty raid" on the women's dormitories (basically a boisterous crowd of beer-swilling males clustering on the lawn outside the

Me in my Army ROTC uniform at the University of Florida, 1962.

windows and daring the girls to show their underwear). By the time I left six years later, university administrators were calling in the Gainesville city police to disperse the masses of angry, rioting students who were protesting the war.

Though the Selective Service had granted a temporary draft exemption for college students, that protection was rescinded for graduates and drop-outs. It looked increasingly likely that my career in aerospace was going to be delayed by a two-year stint as an infantryman. I'd already discovered that a University of Florida engineering degree then took not four, but five years, which had pushed out my chance to join the space race to 1966. But with the draft, I was now looking at 1968, which meant I'd probably get in only at the tail end of it all. Avoiding service wasn't an option, so I decided at least to do it on my terms (and in my hopelessly naïve understanding of what war was like, thinking I'd increase my odds of getting through it in one piece). The University of Florida had a robust Reserve Officers Training Corps (ROTC) program at that time. Every male student was required to take two years of Army or Air Force training, and if you spent four years you could graduate as an officer. At the end of my sophomore year in 1963, that's what I decided to do.

Though I didn't know it at the time, my experiences the following summer would change all that. A high school friend, Henry Hirsch, who was one of my college roommates and working toward a medical degree, thought it would round out my engineering focus to get more involved with real people (boy, was he wrong). At his suggestion, I took a job during the summer break as an orderly in a public hospital in Fort Lauderdale. I was assigned to a ward dedicated to elderly, end-of-life patients, and it was a disaster. Many of those people had no one who came to visit them, and they were both needy of and grateful for any kind or concerned attention. To their credit, the hospital nursing staff there were wonderful in that way. But for me, working with those desperately lonely and ailing older men and women, who looked forward to my just saying hello in the morning, reawakened a distressing self-doubt. It didn't help that my responsibilities included more than just changing sheets. I was also tasked with administering a few medical procedures, like inserting and removing catheters in bedridden male patients (this would only be done by a nurse or physician today). One Monday morning I came in to work to learn that an elderly male patient I'd been caring for daily for almost two months had died over the weekend. He had been in decline before I started work there, yet I was overwhelmed by a troubling, if unfounded, anxiety that somehow I hadn't done enough to prevent his death. I couldn't focus on my job; I second-guessed even my most routine duties. I was a mess. I resigned at the end of that week. Taking off the rest of summer helped, but it was hard to shake off the sudden tumble I'd taken.

I was still a bit frazzled in September 1963 when I returned to start my junior year. I enrolled as planned in Advanced Army ROTC, which would grant me a second lieutenant's commission when I graduated. While there was patriotism involved—I was one of "Kennedy's Children" after all—it was also an act of precaution.

Then in November President Kennedy was assassinated. JFK was my inspiration, the embodiment of a hope-filled future for me, and now he had been murdered. It's hard to convey how that event conspired with my already unsettled confidence to overwhelm me emotionally at that point in my life. My grades faltered, threatening my scholarship. There was only one obligation that forced me to stay in the moment, to get over myself, and that was Army ROTC. The training and drills and discipline demanded of me wouldn't brook inattention or disinterest. If I failed there, I was headed for the draft. By the end of my junior year I had pretty much pulled it together.

During the summer between my junior and senior years I was required to attend Army ROTC officer's training camp at Fort Bragg, North Carolina. It was essentially an abbreviated boot camp, with all of the physical challenges that implies. I drove up there with my close friend from high school, and now a college roommate, Paul Cheries, who had also chosen to graduate as an officer. I never made it through the first day physicals. When asked to fill out a psychological evaluation questionnaire (for example, "Have you experienced significant emotional distress over the last twelve months?"), I checked "yes" in too many boxes. That sent me to a medical officer who served as the base psychiatrist. I explained my difficulty in dealing with my temporary job as a hospital orderly, and chose to omit saying anything about my reaction to JFK's death. I also insisted that it was now all behind me. The doctor asked me if I could be certain it wouldn't recur when I was leading a platoon into battle. It was a horrible prospect to envision; men under my command might die and I'd have to deal with that. It was the hospital all over again. "I don't think so," I said. He didn't look convinced and asked, "But it could happen?" I didn't answer him. He made a note and told me he was going to reject me as an officer candidate.

I was deflated; Army ROTC had helped to restore me, it had improved my self-confidence. I really needed to be a part of it. I told him that I wanted to serve somehow, but I didn't want to go home and then get drafted. He assured me I wouldn't be and he classified me 4-F, unfit for military service. That was the end of my brief military career.

I went back to school in the fall of 1964. Eighteen months later I finished my Bachelor of Science in aerospace engineering and promptly married Barbara Porter, a University of Florida graduate I'd known since high school, who'd just earned her teaching degree. Since my 4-F draft classification meant I now had additional time to chase my dream job, I stayed on for another year to earn a Master of Science with a specialty in high-speed aerodynamics—in other words, the dynamics of a spacecraft's reentry through the Earth's atmosphere. In June 1967 I officially became a rocket scientist. I could join the Moon team now with nearly three years left until the end of the decade!

But the consequences of my revised draft status continued to reverberate. I applied for a job at several aerospace companies, all of which were doing significant hiring in response to the U.S. space program and to the ongoing Cold War with the USSR. With my newly minted degrees, I thought I was an ideal candidate. None of them offered me a job. I came to find out that my 4-F classification, unfit for military service, had scared them off. One company claimed that it would prohibit me from obtaining a U.S. Government Security Clearance, which was required for any classified work they might later assign to me. Just why they thought that is still a mystery, since it subsequently proved to be untrue. All but one company, General Electric, turned me down.

So my first engineering job after graduating from college took me well off course from my plan to work on the Apollo Program. On July 5, 1967, I was hired by General Electric's Missile and Space Division outside Philadelphia, and my new wife and I relocated to an apartment in Audubon, Pennsylvania. GE was under contract to the U.S. Air Force to design a nose cone for an Intercontinental Ballistic Missile (just in case we had to nuke the Soviets). The work was classified, of course, but in spite of the trepidations of my rejecters, my Secret Clearance was quickly approved, in less than a month. My suitability to qualify for classified work never again came up during the rest of my career. Ironically, here I was doing the very kind of work that companies working on the space program had denied me. But they were going to the Moon, and I wasn't.

My first assignment at GE was to work out the detailed mathematical aerodynamics of superheated air flowing around the flat end of something shaped like a bullet that came screaming through the atmosphere at nearly twenty times the speed of sound (a nuclear warhead, for example). It was exhilarating; it certainly provided a robust run-through of my newly earned credentials and was a lot more difficult than any of the classroom problems I'd wrestled with at Florida. Surprisingly, considering my distaste for mathematics, it was fun, because now working the math wasn't the point—it was just one of the utensils in my engineering tool kit. I diligently labored over the problem for several weeks, even spending overtime hours and a couple of Saturdays on it, yet in the end I was able to solve only part of the problem, a sort of first draft. I realized I wasn't going to get any further.

Knowing I'd flunked my first test at work, and prepared to be told so, I took the results to my manager, Lonnie Marshall, an African American with an advanced engineering degree and one of the two or three best managers in my forty-year career. He looked over my work, nodded a little, and to my relief, encouraged me on the progress I'd made. It turned out this was a thorny problem that, as yet, no one had successfully cracked. The strategy was to assign it to a newly hired and eager to please engineer who wouldn't be hampered by not knowing how hard it was. Lonnie was kind enough not to tell me whether I'd done any better than my predecessors.

My time at GE involved the most complex technical work of my career. It even resulted in my first publication in a technical journal. (When I tried to re-read it recently I discovered I could no longer keep up with the math it required.) But by that time it was already 1968 and not what I'd imagined I'd be doing when I'd responded to President Kennedy's call. Then the aftermath of a horrific accident a year earlier and a thousand miles away came around to change my life yet again.

On January 27, 1967 (just as I was finishing up work on my master's thesis), three U.S. astronauts—Gus Grissom, the second American to fly into space; Ed White, the first American to walk in space; and Naval aviator and rookie astronaut Roger Chaffee—burned to death inside their Apollo Command Module during a routine ground test at Cape Canaveral. I was to discover later just how profoundly the tragedy rocked NASA onto its heels. Every aspect of the program was reexamined, reconsidered, redesigned. It was a black period for NASA and its industry partners, and it threatened to delay the launch of Apollo indefinitely. But when the soul searching was completed and NASA finally announced that the first manned flight, now designated Apollo 7, would be rescheduled for October 1968, I realized what I had to do.

Gus Grissom, Ed White, and Roger Chaffee (*left to right*) died in the Apollo 1 Command Module fire, January 27, 1967. NASA Image.

NASA's re-planning in the wake of the horrible calamity that had befallen the Apollo Program gave me a second chance to be a part of it. I rewrote my resume to emphasize my high speed aerodynamics engineering experience at GE, noted that I currently held a U.S. Government Security Clearance, and mailed applications to several companies I thought might be interested. Within a few weeks I got a call from TRW Systems Group in Houston, Texas, a company I knew little about beyond what was in their "engineers wanted" advertisement in *Aviation Week* magazine. They had a contract with NASA for supporting the Apollo Program, and they were located near the Manned Spacecraft Center. Good enough for me. They invited me to fly down for a job interview, which I did, and it must have gone well, because a week or so after I came home, I got a call from Bob Harris, a senior manager there, offering me the position of Member of the Technical Staff, the company title for engineer, on the Apollo Reentry Support Team. Barbara and I packed up again and drove to Houston that October—the same month the manned Apollo launches were scheduled to start. I was determined to earn my keep as a manned space program rocket scientist, and I'd get there in time after all. It would turn out to be quite a ride.

Apollo 1: Failure Was an Option

To understand the "failure is not an option" culture that pervaded all of NASA when I first came to Houston in 1968, it's important to understand what was lost in the Apollo 1 fire on January 27, 1967, and why. That tragedy touched everyone who worked on the program professionally and personally, NASA and contractors alike, and changed forever their attention to the sudden and unthinkable hazards that spaceflight involved.

Interior of the Apollo 1 Command Module after the fire. NASA Image.

On January 28, 1967, one day after the fire, an independent investigation committee was formed, comprising ten aerospace professionals drawn from inside and outside NASA, to determine exactly what had caused the tragedy. When they released their final report on April 5, they concluded that the problems were both technical and, sadly, operational. They found that the tragic fire in the Apollo 1 Command Module was caused by flawed design, flawed engineering, questionable management decision making, and poor maintenance and inspection. In other words, not just faulty equipment but insufficient attention to detail by those in charge killed Gus Grissom, Ed White, and Roger Chaffee. And it could have been prevented—it shouldn't have happened.

NASA took the severe assessment and the committee's recommended changes to heart. In spite of their mandate to land a man on the Moon before the end of the decade, they immediately halted plans for future launches until they could make every change required: to the spacecraft, to procedures, and even to the management within both the government and contractors. When NASA finally announced that the launch of the manned mission, now named Apollo 7, would occur in October 1968, twenty-one months after the Apollo 1 fire, the redesigned Command Module those astronauts would use was a vastly improved model compared to the one in which Grissom, White, and Chaffee died. And beyond that, NASA worked diligently to change their culture from top to bottom of the organization. No longer would one of their own die because someone who was supposed to be supporting them had lost single-minded focus on safety and meticulous attention to every detail.

That effort started immediately. Three days after the Apollo 1 fire Gene Kranz, the Houston Mission Control Flight Director, who would later coordinate the rescue of the crew of Apollo 13, called his team together to give them the directions they would follow for all future missions. In his memoir *Failure Is Not an Option*, published in 2000, Kranz recounted his speech to the controllers that day:

> From this day forward, Flight Control will be known by two words: "Tough and Competent." Tough means we are forever accountable

> for what we do or what we fail to do. We will never again compromise our responsibilities. Every time we walk into Mission Control we will know what we stand for. Competent means we will never take anything for granted. We will never be found short in our knowledge and in our skills. Mission Control will be perfect. When you leave this meeting today you will go to your office and the first thing you will do there is to write "Tough and Competent" on your blackboards. It will never be erased. Each day when you enter the room, these words will remind you of the price paid by Grissom, White, and Chaffee. These words are the price of admission to the ranks of Mission Control.

That was the NASA I encountered when I arrived in Houston in October 1968.

4

LIFTOFF MISSION ONE

CATCHING APOLLO

Clear Lake City: The Cow Pasture Space Center

The division of the TRW Systems Group that hired me was located about twenty-five miles southeast of Houston, Texas. TRW's offices were literally across the street from NASA's Manned Spacecraft Center in Clear Lake City, a newly built community that encircled NASA's property. TRW Corporation has long since divested itself of aerospace interests, but in those days its Systems Group headquarters located in Redondo Beach, California, was a major player. The Houston location that hired me had won a contract with NASA's Mission Planning and Analysis Division to provide support for all phases of the Apollo flights. In practice we were to do a lot of the engineering grunt work preparing NASA for each mission, and also support the NASA staff during the flights.

I started work on October 14, 1968, three days after the first launch of an Apollo spacecraft with humans on board. I was introduced to my new manager, Bob Manders, an aerospace engineer who'd been working there since 1966; was assigned a desk and shown where the restrooms and cafeteria were located; and met my coworkers, while Wally

Apollo 7 Crew (*left to right*): Donn Eisele, Wally Schirra, and Walter Cunningham, 1968. NASA Image.

Schirra, Donn Eisele, and Walt Cunningham orbited overhead. The mission was designated Apollo 7, the first manned space flight of the Apollo Program, a chance for NASA to demonstrate all they'd learned from the Apollo 1 fire and to restart the manned program to meet their end-of-the-decade mission to the Moon.

Bob Manders was, in his words, an Iowa farm boy who ended up sending men to the Moon and getting them back safely. He'd graduated from Iowa State University and was hired by Douglas Aircraft Company's Missile and Space Systems Division in California to do launch and reentry studies of U.S. Air Force missiles. He also started working on a Master of Science degree in aerospace engineering from the University of Southern California. While he was still working on his degree, one of his USC professors, who also worked at the Space Technology Laboratories in Redondo Beach, offered Bob a job. The company, which would later become TRW Systems Group, had won a contract with NASA in

Houston to support the Apollo Program, and they needed engineers. Bob took the job, and his new boss asked him to go to Houston for three months to help define the Apollo reentry corridors. By the time I got there Bob's "temporary assignment" had turned into a much longer one. It would be many years before Bob got back to USC and finished his master's.

I also soon realized just how well my new assignment fit my dreams. At one point I was telling Bob about my interest in science fiction and he revealed that he was a member of the Science Fiction Book Club, which was, and still is, a monthly subscription for SF books. Bob said if I joined on a trial basis through him, he could get a free book. Later I did, and that marked the beginning of my rediscovery of science fiction as an adult.

When Barbara and I arrived in Clear Lake City, it was far from the bustling, hi-tech, fashionable suburb it later became. It was a hot, humid, godforsaken place. When she and I drove around during our first week, exploring neighborhoods searching for someplace affordable to rent, the temperature soared into the nineties every day and the humidity was right up there with it. It was mid-October!

A well-worn bromide passed around by Texas locals, as in several other southern states, assures newcomers that they'll get used to that sort of weather as soon as their blood "thins," whatever that means. If so, my blood must be especially thick. In the sixteen years that I lived in that sweltering crock pot, I never got used to having my glasses fog up when I stepped out of my car in the morning, or having my shirt sweat-pasted to my back by the time I'd crossed the parking lot. There may be advantages to living in that southeast Texas climate, but thinning your blood isn't one of them. Yet, in spite of the stifling heat and humidity, I stayed in Houston for sixteen years for one reason, because that's where NASA was.

The Manned Spacecraft Center had been built four years earlier on a thousand acres of undeveloped pasture located about twenty-five miles southeast of downtown Houston, in the middle of nowhere. That cattle grazing ground was not on the original short list of nine locations chosen by NASA's site selection team to meet a detailed set of criteria established by Congress (access to water transport, a major airport nearby, moderate climate, etc.). Tampa and Jacksonville, Florida, were, as were San Diego and San Francisco, California. Sigh.

For some reason the selection committee reconsidered its list, and this time the Clear Lake cow pasture was added. Possibly the reason was that a Texan, Vice President Lyndon Johnson, was head of President Kennedy's Space Council, and Texas Congressman Sam Rayburn was Speaker of the House, and three other Texas congressmen directly controlled NASA funding, and Rice University had agreed to donate the land (which had been donated to them by Humble Oil Company of Houston). But "moderate climate"? Seriously?

NASA referred to the Manned Spacecraft Center as the Houston Space Center throughout the Apollo Program, though the area wasn't swept up into the city limits until 1980. I suppose the name Cow Pasture Space Center wouldn't have sounded very "space age." In 1973 the NASA facility was renamed the Johnson Space Center.

The 1,600-Mile Roller Coaster

And what, exactly, was my new job I mentioned earlier? My education and GE work experiences were about using mathematics to predict how high-speed spacecraft, or Air Force missiles, would behave when they came roaring back from outer space and plunged into the Earth's atmosphere. That's what Apollo did when it came back from the Moon, except that the Apollo Command Module (the only part of the Apollo spacecraft that actually made it back to Earth), was shaped like an oversized Hershey's Kiss, with the astronauts cramped three abreast into a workspace smaller than the inside of a car and chockablock with equipment. Not quite the sleek, needle-nosed spaceships from my college courses. Another complication was that the Command Module entered Earth's atmosphere fat side first, giving it all the "flight aerodynamics" of a Stonehenge pillar.

My job was to work on a plan for the so-called reentry phase of the mission—the part that starts when the Command Module first reaches the topmost edge of Earth's air and ends when it drops into the ocean. But what was there to plan? Couldn't it belly flop into the atmosphere, which would certainly slow it down, and then descend on

The reference manuals I used to calculate the Apollo reentry paths.

its parachutes until it splashed into the ocean? The short answer is: it wasn't that simple.

First of all, Apollo couldn't splash down just anywhere; it had to make its way across the Pacific Ocean to the U.S. Navy aircraft carrier waiting to pick up the crew. The 12,000-pound flat-bottomed teardrop had to learn how to "fly" to its destination. After leaving the Moon, the Apollo spacecraft fell home "downhill," covering a quarter of a million miles in three days, gaining speed the whole way. By the time it entered the Earth's atmosphere it was traveling at about 25,000 miles per hour, ten times as fast as a modern rifle bullet. At that speed even a boulder can fly.

Let me give you an example. The next time you're cruising in your car at, say, 40 mph, try this experiment, from the passenger seat of course. Lower the window and put out your hand, palm flat, fingers together, and pointed at the sky but tilted forward a little. The wind's force

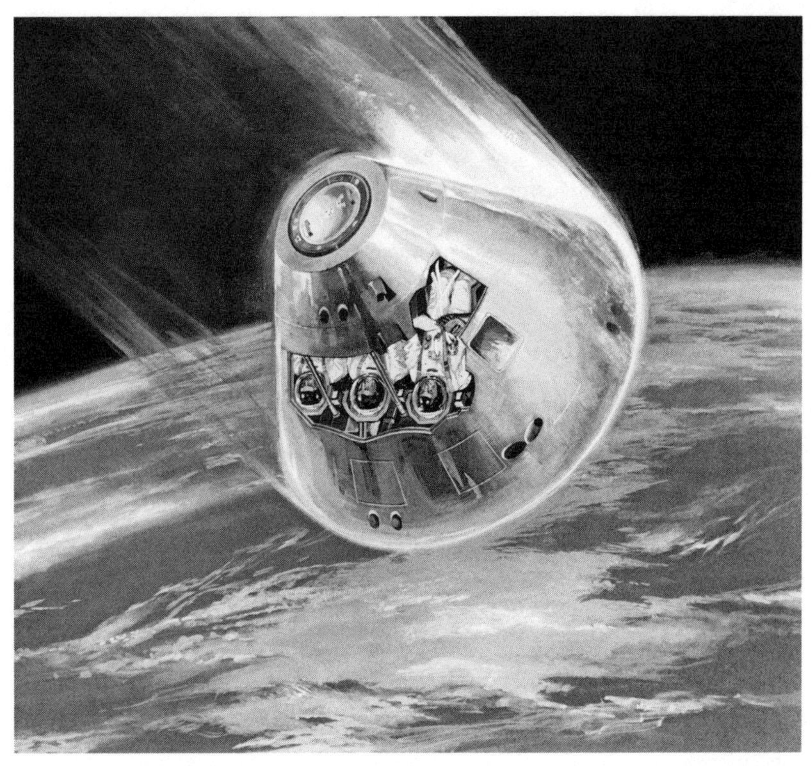

Atmospheric heat buildup during an Apollo Command Module reentry. NASA Illustration.

will jerk your arm upward. Rotate your hand ninety degrees and it'll kick to the side. (Watch out for drivers in the oncoming lane who may not be as interested in science.) Rotate your palm down and your arm will dive toward the road. You get the idea. That's how the Command Module flew. It was designed with its weight a bit off-center, so that when it came back into the atmosphere the flat side would tilt about twenty degrees into the direction of the wind, if that's the right word for the air coming at you at thirty times the speed of sound (a velocity that aerodynamics folks call a Mach number of 30).

Evenly spaced around the tilted wide end of the spacecraft was a set of small rockets that faced crosswind. Short bursts from these jets caused the Command Module to roll on its long axis, much as your arm might twist if you rotated your palm outside the window. If the spacecraft isn't slowing down quickly enough, the jets fire to rotate the tilt downward and push the Command Module deeper into the atmosphere. Slowing down too quickly? Rotate up again and loft like a stone skipping over a pond. Need to get centered on a path to the recovery ship? Rotate partway to the left or right and move sideways. An Apollo reentry was a fifteen-minute, 1,600-mile, high g-force, gyrating roller coaster ride: not for the faint of heart. Part of my job was to help teach the astronauts to fly the Command Module, which takes me back to the bigger picture.

The complete Apollo spacecraft system was a complicated and many-faceted machine. Its Command Module (CM) housed the astronauts and served as their command center during a mission. The Service Module (SM) was unmanned and provided power, water, and oxygen to the Command Module and also contained a rocket-based propulsion system. The Lunar Module (LM) was a fully autonomous powered spacecraft designed solely to ferry two astronauts to the Moon's surface and return them to lunar orbit for rendezvous with the Command Module once the surface exploration phase was complete.

At liftoff, these three modules rested atop a massive Saturn V rocket capable of generating 7.6 million pounds of thrust. No element of that Apollo system was intended to be reused. As a mission progressed, its components were sequentially discarded into the ocean, or into space, or abandoned on the lunar surface. The total launch weight of the assembly was more than six million pounds, but only the 12,000-pound Command Module returned to Earth. It splashed down in the ocean and was recovered, but it was not reused.

The Command Module had a small computer onboard to perform a variety of tasks, including automatically controlling the reentry flight. It was an autopilot designed to guide the craft on a sinuous, high-speed trajectory to a safe ocean splashdown. Both the computer itself and the programs that ran on it were newly developed for the Apollo Program. No previous craft of any description had been called upon to make such a "flight." With an excess of caution that characterized the NASA culture after the Apollo 1 fire, each phase of an Apollo mission required a backup plan—an alternative means for the astronauts to complete that segment using only redundant instruments and manual controls; in other words, independent of the autopilot.

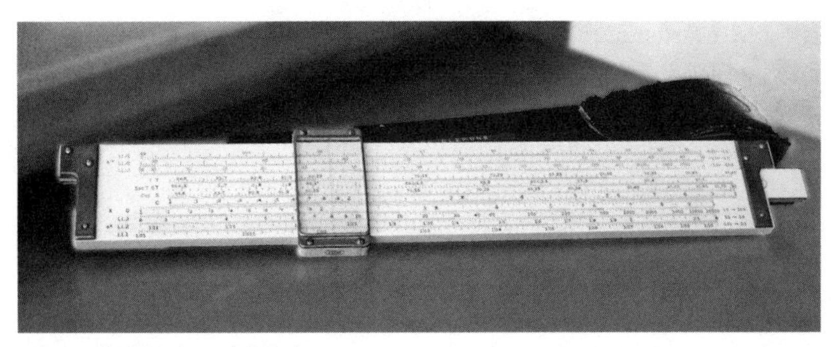

My Apollo Program slide rule.

Based on my recent degrees in engineering, and my one year of experience designing a missile that sped through the atmosphere at white-hot speeds, I'd been assigned to the TRW team working on the reentry of the Command Module. I was twenty-four years old at the time: young, but not so very. The average age of the NASA team working in the Mission Control Room in those days, the professionals TV viewers saw during an Apollo mission, was just twenty-eight. The specific task of the team I joined was to develop a backup process for the astronauts to use during reentry in case of a computer failure, and then to test it. Just how were we to do that?

As I write this, I'm sitting in an armchair using Microsoft Word on my Dell laptop that's resting on my . . . um . . . lap. I click a calculator icon to convert Mach number to miles per hour. Next, I leap over to Google search some factoid I would otherwise have to rummage through old storage boxes to find. It's hard these days to focus with any clarity on how I or anyone else did engineering work on the Apollo Program back in the dark ages of 1968. How primitive were we then?

The first hand-held calculator, the Hewlett Packard HP-35 Slide Rule Pocket Calculator, didn't come on the market until 1972, three years after Neil Armstrong walked on the Moon. It eventually replaced my slide rule and the TRW engineering department's clattering mechanical calculator, even though it went for a hefty $395 in 1969 dollars—about the same as a twenty-four-inch color TV back then. That's around $2,200 today (I just Googled that). I remember a colleague bringing his newly purchased HP-35 into the office one Monday morning. The rest of us gathered around and pawed at it and ooh'd and aah'd, privately wondering if he was just crazy or if his rich uncle had died.

That was in 1972, by which time I had pretty much wrapped up my work on Apollo—the last Moon landing was that December—so the Hewlett-Packard pocket calculator came along too late for Apollo. I did make heavy use of it working on early Space Shuttle studies after TRW bought a few of them for us to share (as long as we locked them up at the office each night).

So, yes, we went to the Moon using pencil and paper and a slide rule. We were neither naïve nor deluded. We had been challenged by a young, charismatic president to realize an ages-old dream of mankind. We believed we could take advantage of the new technologies we had, and then invent the rest to accomplish it. We were adventurous but not foolhardy. We laboriously calculated the risks, and even more laboriously worked out ways to reduce them. What we lacked in computer resources we made up for in tedious, demanding work and in seemingly endless planning, training, simulation, and rehearsal. And we did it!

The first part of my assignment was to understand exactly how the Command Module's onboard computer was supposed to work under these "flight" conditions. Next, I needed to learn to distinguish between the behavior of a correctly operating autopilot under a wide range of reentry conditions and any erroneous commands it might issue if something went wrong. Finally, in case the autopilot failed, I had to develop the procedures an astronaut could use to take over control and manually pilot the spacecraft to a successful splashdown.

The work I'd done designing a missile nosecone for GE was straightforward compared to the atmospheric flight of an Apollo Command Module. There were a few "old hands" in the office (perhaps thirty years old) who had done some of the preliminary work on developing crew procedures for reentry. They were a capable group of professionals, but they were not specialists who had experience with extremely high speed aerodynamics. My boss, Bob Manders, took me aside and told me that NASA wasn't happy with the results TRW had produced thus far. He said I should look over the work my colleagues had done, but I might have to start over. Meanwhile the early Apollo mission crews would be training, and flying, with a version of our procedures that he'd heard the astronauts considered pretty much useless. I found out Bob wasn't exaggerating. Start over? I didn't even know the basics. I had to catch up quickly and figure them out. My team leader, Dennis Zelk, had been slugging through a lot of this analysis alone and he was more than agreeable to help me. That's how I spent my first few months.

Apollo 8: "There Is a Santa Claus"

The year 1968 had been a traumatic one for the nation, and I wasn't immune to its effects. Though JFK was assassinated in 1963, my wound from that was still raw. His vision was the reason I'd made rocket science my career, so even five years later I couldn't watch video footage of him without tearing up. Many of my friends and classmates, including my best friend from high school and college, Paul Cheries, were even then slogging their way through the jungles of Vietnam. If not for my draft status, I would have been there beside him. The Tet Offensive launched by the Viet Cong early that year exposed the reality that we were caught in a brutal and seemingly endless war we could never win. In April Martin Luther King was assassinated, and in June so was Bobby Kennedy, another early inspirer. I was still living near Philadelphia when Robert Kennedy was killed and grieved as I watched his funeral train pass through on its way to Arlington Cemetery. While changing my career to the Apollo Program and moving to Houston late that year was exciting, it was shadowed by an abiding sadness and lingering uncertainty. The days of Camelot were well behind us.

So when NASA launched the Apollo 8 mission on December 21, 1968, even though I had little input into the planning for that flight—barely two months into my job—I already felt a sense of ownership, of accountability, for making this first critical step toward achieving President Kennedy's mandate. On December 24, 1968, at 4:03 a.m. Central Standard time, Apollo 8 reached the Moon and completed its lunar orbit insertion burn—the rocket firing that placed it into a stable orbit that circled Earth's distant and barren satellite—making astronauts Frank Borman, Jim Lovell, and Bill Anders the first human beings to visit another world. The success of that maneuver was vital; a misfire

Apollo 8 Crew (*left to right*): James A. Lovell Jr., William A. Anders, and Frank F. Borman II, 1968. NASA Image.

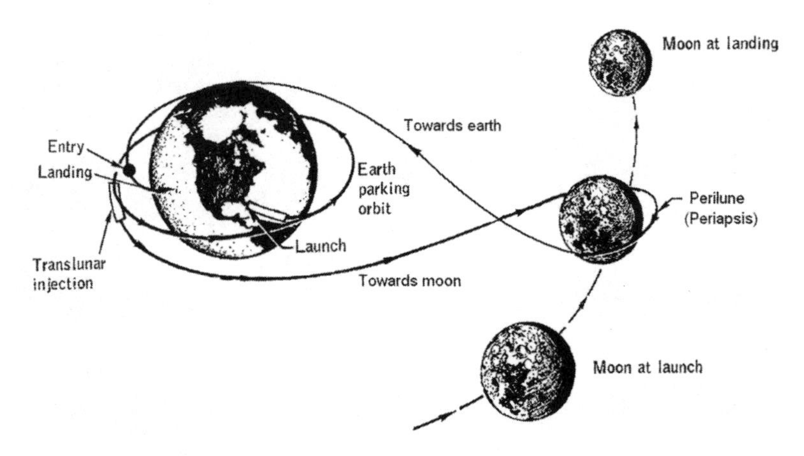

Backup "free return" trajectory for Apollo 8 during its flight to the Moon. NASA Illustration.

could have sent the spacecraft either spinning off into space or crashing into the lunar surface. I got up early to watch the TV coverage from home. When Jim Lovell's voice radioed that the burn was complete, I, like millions of others, sighed with relief.

Apollo 8 spent twenty hours in lunar space, completing ten orbits around the Moon. The astronauts were the first to view the lunar far side, and the first to witness the distant Earth rise above the Moon's horizon. During one of their orbits they made a live television broadcast back to Earth and read the opening lines of the Book of Genesis. News network estimates at the time confirmed that it was the most watched TV broadcast in history.

They needed one more successful rocket firing, the so-called trans-earth injection burn, and they'd be on their way home. Like the one that had put them into orbit, the burn to get them back out of it was just as critical. If it didn't go as planned, Apollo 8 could orbit the Moon forever. NASA might have had to repurpose the crews of the next Apollo missions to retrieve the bodies of their fellow astronauts.

The engine's firing was scheduled for around midnight on Christmas Eve, but I wasn't watching TV then. As was our custom in those days, my wife and I were attending midnight mass at our local church.

The residents in the neighborhoods surrounding the NASA Manned Spacecraft Center were almost entirely people who either worked on the Apollo Program or serviced the daily needs of those who did. At least two Apollo astronauts, Jim McDivitt and Gene Cernan, were members of our parish. The Clear Lake City clergy of every denomination made it their business to stay well-versed on the essentials, and the risks, of manned spaceflight.

As midnight approached I fidgeted in the pew and kept glancing at my watch, impatient with the more elaborate and protracted rituals associated with the holiday. I wished I had gone to one of the morning services so I could be home watching the news coverage of the flight. At around 12:30 a.m. or so, an altar boy, who had evidently been stationed just beyond the side door, stepped into the sanctuary, walked over to the pastor who was saying the mass, and whispered something to him. The priest paused to listen, nodded a couple of times and then turned around to face us.

"The trans-earth injection burn was successful," he announced. "Apollo 8 is on its way home."

As we absorbed the news there was first silence, then a collective and audible sigh, and finally a loud and uninhibited cheer. Parishioners stood up and clapped and shook hands and embraced the person nearest them. When at last we settled down again, the pastor added, "And the crew of Apollo 8 wants you to know that there is a Santa Claus." That night I became a member of the space program community.

Goldilocks and the Three Astronauts

In spite of my newfound pride at being an "insider," at work I was still struggling to understand exactly how the Command Module controlled its reentry through the atmosphere. I'd spent my first month or so crawling through all the NASA documentation I could get my hands on, and I spent hours in meetings with Dennis and my other colleagues talking me through the earlier procedures they'd developed. Gradually I began to see why the approach they had taken was too simplified to work during extreme supersonic flight through a real atmosphere. I understood then that my background at GE in high-speed aerodynamics was the very reason my boss had hired me. He also assigned another young aero engineer named Phil Moseley to the team.

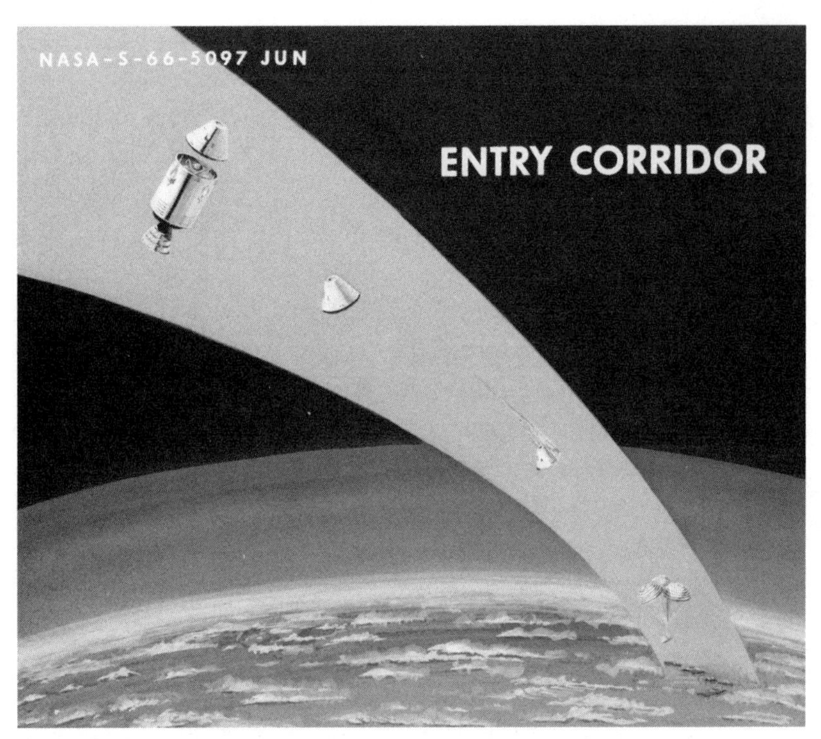

Apollo Command Module reentry corridor. NASA Illustration.

At the same time Phil was working with me on Apollo, he was also spending nights and weekends getting a Doctor of Mechanical Engineering degree at Rice University. (After he earned his PhD, in 1972 Phil left TRW to work for Exxon helping to design the computer programs they used to simulate their production process. In 1982 he formed his own consulting company named P. E. Moseley and Associates, which today is a leader in software and services to the petroleum industry.)

Phil was a brilliant guy. I was lucky to be able to work beside him on this problem because unfortunately, our boss was right: we had to start over. The problem the astronauts were having with TRW's earlier backup procedures was that they were too rudimentary. It was as if you could use your car's navigation system to drive a direct route to your house, but if you took a detour or went by any different roads, you had to figure it out for yourself, because your car's instruction manual said you shouldn't trust the computer doing that. That was problematic, since an Apollo reentry never took the same route twice.

To explain how we prepared for an Apollo reentry in the 1960s with the tools of the age, I need to go into a little detail about the technology we used. Before I could develop a workable backup plan for the astronauts to use, I had to answer two questions. First, how would the onboard computer autopilot behave when it was correctly controlling the Command Module during reentry? Second, how would it behave if a glitch or software bug cropped up? In other words, how could an astronaut differentiate between acceptable and erroneous computer commands during the fifteen-minute roller coaster reentry ride? What was the range of possible conditions the computer could handle? For that I had to deal with some what-ifs.

What if the Command Module reentered the atmosphere at a different location over the Earth than had been planned? For example, if the astronauts spent more or less time coming back from the Moon than they'd scheduled, because of the Earth's rotation they would enter at a different longitude and latitude. This would mean that the distance and direction to the recovery ship would be different, which would result in a different reentry ride. Later on, Apollo 13 gave us a dramatic example of that.

There was also a Goldilocks problem. The planning for Apollo's reentry begins at the Moon. When the Service Module engine burns to leave lunar orbit, it places the spacecraft on a long arcing path, called a trajectory, which intersects Earth's atmosphere at a very specific angle. What if that trajectory intersected the atmosphere at a different angle than was planned? One way that could happen was if the SM engine burn was longer or shorter than expected. For an object moving at thirty times the speed of sound, compressing the surrounding air enough to raise its temperature to 5000°F (hot enough to melt iron), the atmosphere packs quite a wallop. How big a wallop depends on how the object makes its reentry.

To use a person diving into a swimming pool as an example, there's a big difference—to your body, not just your ego—between an arms-first sluice from poolside, Michael Phelps style, and a belly flop off the high diving board. If the Apollo trajectory intersected Earth's atmosphere at too steep an angle—that is, if the craft "belly flopped"—the stress from impact, called g-force or g's (a multiple of its weight), could be great enough to do serious damage. Although the CM structure itself could withstand well over 20 g's, a human experiencing that level of force for an extended period could suffer severe injury or even death. Even at lower g's, a person's ability to perform even ordinary tasks is severely impaired.

On the other hand, a trajectory that intersects the atmosphere at too shallow an angle produces an equally dangerous problem. During reentry the Command Module has no rearward facing engine to slow it down; it relies on a dive into the thickening atmosphere to do that. If that dive is too shallow, the spacecraft will pass through the thin upper reaches and swing back into space. Once the Command Module frees itself from the Service Module in preparation for reentry, it operates solely on internal batteries that have a life of about three hours if everything is turned on. Even if the CM stayed in Earth orbit, the battery power would drain severely before the astronauts could cycle back around, reenter, and land.

The acceptable range of angles for Apollo's trajectory to intersect Earth's atmosphere was called the reentry corridor. That corridor was pretty narrow: enter no shallower than about 5.25 degrees below horizontal, and no steeper than about 7.5 degrees. In addition to everything else, because of the extreme speeds, a reentry flight within that corridor still had quite a number of possible outcomes. The g-force could range from 1 g to 7 g's, and the spacecraft's guidance system could fly it downrange anywhere from about 1,300 to 2,200 nautical miles and crossrange up to 80 nautical miles. That left an area of around 40,000 square miles of ocean as a landing area, a definite challenge for positioning the U.S. Navy recovery ships.

So back to my problem. When was it safe to allow the computer to control the return? How was the onboard autopilot supposed to operate during reentry? That meant digging through the inner workings of the Apollo Guidance Computer itself.

Learning the "Ropes"

The Apollo Guidance Computer, which housed the reentry autopilot program, was about the size of a small toolbox, twenty-four inches long by twelve and a half inches wide by six and a half inches deep, and could handle no more than 38,000 "words" of "fixed" memory (computer content that could not be modified during a mission) and 2,000 words of so-called addressable memory, which an astronaut could change during flight. That's little more than half the capacity of the first Commodore 64 home computer that came on the market in 1982. Looking at it another way, the Apollo Guidance Computer's memory would handle only about twenty-five pages of a document written in Microsoft Word, and only the first two pages of that could be rewritten.

The astronauts communicated with the computer via an onboard display and keyboard device (DSKY, pronounced "diskee"), which was built into the cockpit. Data entry onboard was primitive by today's standards: a calculator-like keypad (the first of its kind); three registers that displayed rapidly changing five-digit numbers, with no information

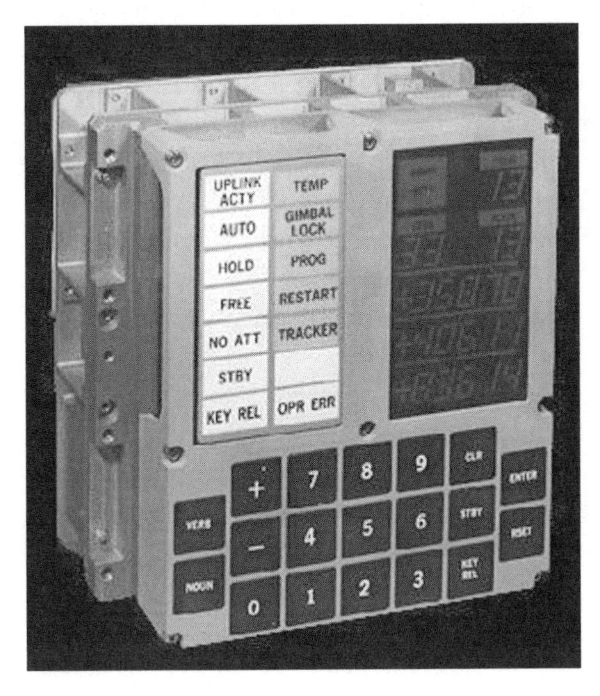

Apollo Command Module onboard Display and Keyboard device (DSKY). NASA Image.

about their meaning; and three displays that showed codes indicating what those current register numbers meant, four simple function keys, and a series of indicator lights. All input to the computer was entered manually by the astronauts via predetermined sequences of keystrokes.

The Apollo guidance software programs ("the autopilot") were coded onto circuit boards mounted inside the computer, one of the first uses of integrated circuits for flight. The programs were designed by engineers at the Charles Stark Draper Laboratory of MIT in Cambridge, Massachusetts, and then "written" by interlacing together thousands of copper wires in precise ways. Wires were woven through tiny donut-shaped nickel-iron cores: a wire passing through a core representing "1" and a wire passing outside a core representing "0." Each core woven in a specific way embodied a single programming step. The cores and wires were wrapped, sealed in plastic, and mounted on a board. A single board of fully intertwined wires and cores, referred to as a "rope," was one program; for example, the reentry guidance program.

Up to thirteen primary software programs, such as leaving lunar orbit or flying a reentry, were required for each Apollo mission. A subprogram performing a specific function, such as calculating vehicle speed and orientation, or firing engines to enter lunar orbit, or performing star tracking for navigation, was then activated by the astronauts, depending on the needs of the mission. The actual weaving of the "ropes" was performed by women working in specialized factories, inspiring some MIT programmers to nickname the product LOL memory, not meaning "laugh out loud" as in today's e-parlance but rather "little old lady" memory.

Once woven, these programs could not be changed during the mission—they could only be initialized with flight-specific data values entered via the DSKY. That simplified my problem somewhat, as I didn't need to worry about missing recent updates to the program. But it also limited the types and amounts of data the astronauts could enter during a mission, and therefore the data available for me to use in developing procedures to monitor the health of the program during a reentry flight.

With limited computer resources onboard Apollo, the real "heavy lifting" during a mission to the Moon had to be done on the ground,

NASA Manned Spaceflight Center Real Time Computer Complex (RTCC). NASA Image.

in Houston's Mission Control Center. The large Apollo Mission Operations Control Room familiar to television viewers, filled with live displays and flickering consoles and headset-wearing flight controllers, was the visible tip of the support team iceberg. Behind the walls of large displays was a warren of smaller support rooms where specialty NASA and contractor teams worked to provide data and analyses requested by the flight controller in charge. On the floor below the Mission Operations Control Room was the Real Time Computer Complex (RTCC), which housed five large computers built by IBM. In 1968 these computers, commonly called mainframes, were IBM System 360s (an early version of which now resides in the Smithsonian Institution) and could operate at a then breathtaking speed of over 100,000 calculations per second. By comparison, today's fastest supercomputers are capable of performing 10.5 quadrillion calculations per second. You could run the IBM 360 Apollo support programs today on a Sony PlayStation while also playing *World of Warcraft*.

Nevertheless, these computers were powerful enough to simulate all phases of an Apollo mission during a flight, including the Command Module's fiery return through the Earth's atmosphere. The primary role

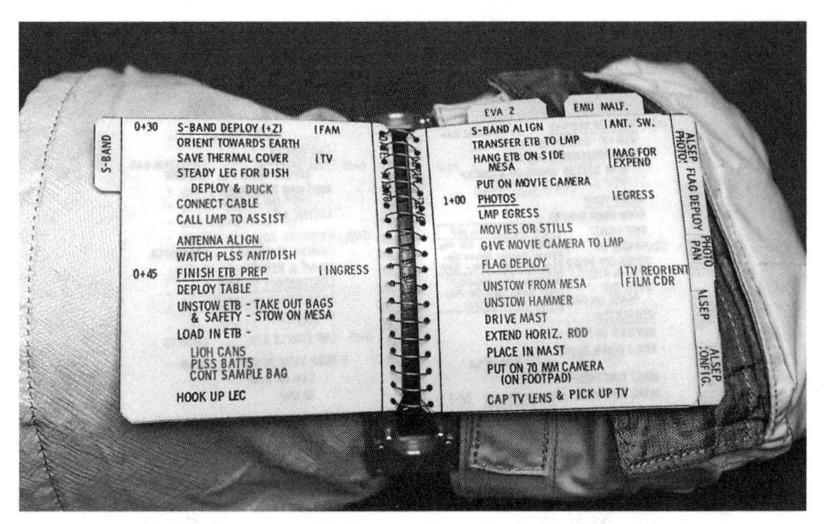

Apollo astronaut's onboard checklist notebook. NASA Image.

of the RTCC was to perform all the calculations needed to monitor and control the health of the Apollo spacecraft from the ground, and to compute all the data and commands required by the astronauts to manage the mission from inside the spacecraft. This included many thousands of parameters that had to be entered by hand into the Apollo Guidance Computer during flight. This was crucial yet tedious work, and a mistake could prove costly. Since the available memory onboard was so small, a new set of data needed to be entered via the DSKY during each mission phase.

The astronauts carried into space a set of spiral-bound notebooks that held pages and pages of step-by-step procedures to follow for each operation. These included pages devoted to each unique set of keystroke entries. On the ground, Mission Control would calculate the specific numerical values to be used; the numbers would be voiced up one digit at a time to a crewman, who would record the digits in a corresponding table in the notebook. The astronaut would read back the entries to Mission Control, who confirmed the numbers. All this had to be done before the data were entered into the onboard computer. As many as 10,500 individual keystrokes were required to complete a single lunar mission. There wasn't much time left over to stare out through the window at earthrise.

Not all the calculations required to plan and support a flight could be done after liftoff, so between missions, engineers like me could also get access to the RTCC computers. NASA had a set of sophisticated computer programs that ran on those IBM computers in the RTCC and that, among other tasks, could faithfully simulate the Apollo spacecraft, its onboard computer operations, and the surrounding external environment as it performed a high-speed reentry through Earth's atmosphere. The programs were designed to allow a wide range of external and internal conditions to be tested and altered without making coding changes to the complex software itself. We could examine the variations for a reentry by inputting factors that could affect a flight, and we could run short standalone programs that worked with the main RTCC programs to exercise them in various ways. In other words, we could make the RTCC computer do what we wanted without altering the essentials of its programming.

So I sat at my desk poring over NASA documents issued for an upcoming mission. These described the planned reentry date and time and location above the Earth as well as planned positions of the recovery ships. Then I turned to tables the U.S. government published encompassing pages and pages of atmospheric data (pressure, temperature, density, and speed of sound, at a wide range of altitudes) for various latitudes and longitudes and times of the year. For example, the air in summer is less dense, thinner, than the air in winter, which would affect how the Command Module passed through that atmosphere during reentry.

I might also need to pull out my slide rule, or run some calculations on our Friden electrical mechanical calculator (it looked like an oversized typewriter with one hundred keys and it could do basic math), or refer to my *TRW Space Data Handbook*, my college textbook on orbital parameters, or my *Handbook of Chemistry and Physics*, which to this day is an aerospace engineer's bible. As I write this I'm looking at the current version, the 98th Edition, 2017–2018. The one I used during my time on Apollo was the 41st Edition, 1959–1960.

Once I worked out a combination of conditions I wanted to throw at the Command Module and its Onboard Guidance Computer (via a surrogate running in the RTCC), I'd set up a number of short programs written in FORTRAN, an early scientific computer programming language invented by IBM, that would put the system through its paces. First I'd write one that modeled everything NASA was planning for on the mission, the so-called nominal run, so I could have a baseline to compare to the others. I've always found NASA's use of the word "nominal" a bit of an understatement. A common definition of the word *nominal* is "small" or "insignificant," as in "a nominal sum of money." But when NASA uses it, the word means "perfect," which means that everything worked exactly according to plan.

Then I'd try other variations, the "off-nominal" circumstances, like changing the entry angle, or the geographic location of the reentry above the Earth, or the distance to the recovery ship, or even the calendar date. Or combinations of these factors. Any of these could happen if a launch was scrubbed, or if Apollo's stay at the Moon was different than scheduled, or if the Service Module engine's burn was longer or shorter, or if the weather wasn't what was expected, or if any number of other unpredicted situations occurred.

Each of these had effects on how Apollo encountered the atmosphere during reentry, and this affected the response of the onboard autopilot while correcting for them. I had to simulate as many of these as I could get RTCC computer time to process, to ensure that I'd captured a comprehensive range of possible reentries the onboard computer might be called upon to fly. Or, using my earlier example, to test the alternate routes your car's navigation system might use in getting you safely to your house.

Once I'd written my set of FORTRAN programs—yes, using paper and a pencil—to model an array of Apollo reentry situations, I sat down at a punch-card machine and typed them in. This device was basically a desk with a built-in typewriter keyboard. Instead of typing words onto paper it punched holes into rectangular cardboard cards, with a specific pattern of holes representing the corresponding typed character, which could then be read by a computer input device at NASA's RTCC.

Each card, about three inches wide by seven inches long, held one line of program code, so that even simple programs could run to tens of dozens of cards, called decks. The punch card machine printed the typed characters above each set of holes, so I could double check the programming embodied in the deck before I ran it. Since our office was across the street from the Manned Spacecraft Center, which housed the RTCC computers, I'd gather up several sets of decks, hundreds of cards, wrap each deck in a rubber band, and load them into a long metal tray. Members of our team took turns carrying the trays over to NASA.

On more than one occasion I dropped a deck I was fiddling with, producing a disheartening game of 52 pick-up. I'd have to spend a half hour scuttling around the hall gathering up the cards and re-sorting them. I was hardly the first clumsy programmer, since the punch machine anticipated my ineptness by printing a sequence number in a corner of each card.

TRW was one of several companies in the area doing studies for Apollo, so the demand for computer time was high. Unless we'd been given a special priority, usually only granted to support an Apollo mission in progress, I'd get the results returned in one or two days. They'd come back printed on accordion-folded green and white striped paper. I'd sit at my desk analyzing the results, or discovering my typing mistakes, and then repeat the process with another set of reentry conditions. The idea was to uncover and document the broad range of appropriate behavior for the onboard computer during a typical reentry, so that the astronauts could know what to expect when all was going well. It was a tedious and labor-intensive way to grind out data, but at the time it was cutting edge, high art, and actually great fun.

Playing Command Module Pilot

Knowing how the Apollo Command Module onboard computer was supposed to work during reentry was only half the challenge; the astronauts also needed to recognize its behavior when it wasn't working correctly. More important, if that did happen, they needed to know what to do about it. Simulating those conditions required entirely different equipment, and there was nothing tedious about that part of my job.

NASA's Manned Spacecraft Center also had more than a dozen computer-driven training simulators that the Apollo astronauts used while preparing for their missions. These ranged from a Lunar Module simulator, which reproduced out-of-the-window views of the Moon's surface as the crew prepared for the lunar descent, to a Command Module simulator called the Flight Acceleration Facility, which rolled and swiveled on the end of a centrifuge and re-created the g-forces of ascent and reentry (much like the Mission Space ride at Walt Disney World in Orlando). Astronauts spent tens of thousands of hours on these machines readying themselves for a mission that could be reproduced nowhere on Earth—a flight to the Moon.

Another one of these machines was a stationary Command Module simulator designed to train the astronauts on the function and operation of the Apollo Guidance Computer. Our team could get access to it when it wasn't otherwise being used, usually only late at night or on weekends. We'd sit in a curtained cockpit and operate a Command Module console outfitted with a complete and comprehensive set of fully functional cockpit displays, lights, switches, meters, and controls as well as an operational onboard computer. This simulator didn't spin—that one was reserved for the astronauts—but it was otherwise quite realistic. It was a bit like driving a racecar arcade game at Dave & Buster's.

If it was my turn to be Command Module Pilot, I'd sit in the lefthand seat, as in an airplane cockpit. From that position I could monitor g-force, elapsed time, speed, and distance to landing. I could scan the gauges and meters and rotating attitude indicators in the mockup spacecraft, while my simulated reentry flight "rolled" and "pitched" and "bucked" its way through the atmosphere. I could watch the autopilot display its interpretation of what was going on, called the "state vector" in NASA-speak, and the commands it issued to control the flight. I could check the backup displays and readouts, which provided information about the flight's progress from instruments and gauges that were independent of what the autopilot thought was going on. I could use the Command Module's controls to override the autopilot and, using the backup system displays, take command of the reentry myself. Flying it was a rush.

The Command Module training simulator could reproduce many of the environmental variables that affected a normal reentry. From the perspective of the astronaut, we could watch the autopilot run through its paces and figure out which displays and readouts the astronauts should check to monitor how well it was doing. Most important, the set-up of the simulator's external controls also allowed us to change the behavior of the autopilot program itself; that is, to introduce errors into its inputs or outputs or to cause it to degrade or even fail in obvious or subtle ways. Each one of our team members took turns putting the Apollo computer through its paces.

I alternated doing much of my "flying" in the simulator with Phil Moseley. One of us sat inside and flew the reentry while the other worked the outside controls to set up the initial conditions or to simulate errors during flight. One reentry might be nominal; that is, the baseline plan. Another could be widely divergent from nominal, perhaps a different entry point, or a longer range to splashdown, or a higher initial velocity, but still within the bounds of what a correctly functioning autopilot could handle. The next might introduce an error that would cause something to go wrong during the flight.

The key was that the one sitting in the cockpit didn't know which conditions his teammate had thrown at him. We had to figure it out by watching events unfold on the gauges, by using the DSKY to call up additional information from the autopilot, by monitoring the backup displays to see if they were in agreement with the onboard computer, and by learning to distinguish between an acceptable reentry autopilot command and a faulty one.

I don't remember how many flights Phil or other members of our team piloted, but during my own turns in the cockpit I ran through dozens of simulated reentries playing Command Module Pilot. Each of us had to build up enough experience to recognize and respond to simulated errors, to take control when one occurred, to use the instruments and readings that could be trusted to fly the remainder of the reentry manually, to make a safe ocean landing, and to make sure that landing was close to one of the Navy ships waiting to pick up the crew.

Now and then my competitive inner demon took charge. Since there was no simulated g-force to contend with, if my reentry was coming in too fast or too far and I hadn't realized that soon enough, I would, on occasion, roll the Command Module tilt down and take a 20-g dive to avoid overshooting the recovery ship, a maneuver that would have been fatal in an actual flight. We didn't include that option in our briefings to the crew. The number of displays, instruments, indicators, and controls involved in flying a reentry was formidable, and yet only a fraction of what an Apollo astronaut had to master to prepare for an actual lunar mission. It made me appreciate the incredible capability and dedication of those chosen to go to the Moon.

When our reentry planning team finally worked up all of our numbers, and ran our computer programs, and analyzed our data, and recorded what we'd learned in the Command Module Simulator, we documented the results, complete with charts and graphs and tables of numbers, as "Recommended Entry Monitoring and Backup Control Procedures" for the upcoming mission. These were major revisions to the backup procedures TRW had produced earlier, and they were significantly more detailed and complex. We briefed our recommendations to our NASA counterparts, and sometimes to the astronauts who were assigned to that mission. We explained what we'd done, how we came up with our revised procedures and how to use them, and also how to use the new charts and graphs we'd designed for the astronauts to carry along and refer to during their flight. I was pretty proud of myself by then. Barely three years out of college, and only six months into my job with TRW, and here I was teaching the Apollo astronauts how to pilot their spacecraft. I'd be educated later on how little I knew.

Apollo 9: An Unearthly Spider

The Apollo Moon Program was my introduction to large and very complex national programs. In later years I worked on NASA's Skylab and Space Shuttle programs and, after that, on the FAA's Air Traffic Control Systems modernization. In industry parlance these projects are called "mission critical," meaning that strategic national goals and human lives depend on their successful operation. But Apollo was something more. Even in the middle of it all, those of us working on it understood we were doing something special, something unique in human history. We were realizing a human dream that had begun with the dawn of man. It was overwhelming then, and it's overwhelming now.

With the successful checkout of the spacecraft in Earth orbit on Apollo 7 and the daring flight to the Moon and back on Apollo 8, by

Apollo 9 Crew (*left to right*): James A. McDivitt, David R. Scott, and Russell L. Schweickart, 1969. NASA Image.

the end of 1968 NASA managers and executives believed they had the majority of the data and expertise needed to press ahead with a schedule for a manned lunar landing by the end of 1969. They established a breathtaking pace to do so. In March 1969 Apollo 9 would flight test the newly completed Lunar Module spacecraft in Earth orbit. In May, just two months later, Apollo 10 would take the full set of spacecraft to the Moon and perform a "dry run" of the lunar landing, stopping just short of actually touching down. In July Apollo 11 would attempt to land on the Moon and have a man leave the spacecraft and walk on its surface. Just in case that flight hit a snag, in November, Apollo 12 would make a second attempt to do so. The United States would go from having its first manned test flight of an Apollo spacecraft in October of one year to landing men on the Moon nine months later, just about the time it would take for a newly conceived baby to be born.

Every one of those flights needed a specialized plan for reentry. Because the Apollo 8 mission had been redirected to the Moon instead of Earth not long before I arrived, some significant revisions had to be made to its entry planning, and I'd contributed a bit to that. But with the planning for Apollo 9 my learning curve took a sharp turn upward.

As ambitious, successful, and in retrospect risky, as the flight of Apollo 8 had been, no one was going to walk on the Moon without fully testing a completely new type of spacecraft with a revolutionary design that would allow them to do so. The Apollo Lunar Module (LM, pronounced "lem") was a spacecraft unlike any that came before it or that had been imagined. Early on, the vehicle was called the Lunar Excursion Module, but NASA management didn't think the word *excursion* was appropriate for a Moon landing, so the middle word was dropped; but the pronunciation persisted.

The Lunar Module was the first self-contained and self-propelled spaceship designed for human use solely in the weightlessness of space and on the surface of the Moon. It could not fly in the atmosphere of Earth and would not survive reentry. It was an incredibly complicated machine. It had a six-by-six-by-six-foot two-person crew compartment, standing room only, with small triangular windows mounted into a wraparound aluminum skin about as thick as that of a soda can. It had two entrances (or exits), one overhead so that the crew could move between the CM and LM, and one forward, between the triangular windows, to allow access to the lunar surface. In addition to the astronauts themselves, and the gauges, controls, and apparatus built into the LM, that cramped space had to hold food and supplies, sleeping hammocks, spacesuits, tools, cameras, and other equipment for use on the Moon and, on the way back to Earth, to safeguard the moon rocks that the astronauts had gathered. Imagine jamming all of that plus yourself into a small closet.

But there's more. For a landing on the Moon, two of the Apollo astronauts descended to the surface in the Lunar Module while the third crew member stayed aloft in the Command Module, continually orbiting overhead. The Lunar Module itself was actually two spaceships in one. It had distinct and separable components: an ascent stage that included the crew compartment, and a descent stage upon which the ascent stage rested. That bottom stage was basically a flat octagonal platform with four deployable struts and a 10,000-pound thrust landing engine. The rocket provided the power required for the LM to descend from orbit and touch down. Later, when the Moon landing astronauts were ready to leave the lunar surface, the descent stage platform became their launch pad. They ignited a second, smaller engine built into the ascent stage, exploded the latches connecting the two stages, and lifted off, leaving the bottom stage behind. The ascent engine, with its 3,500 pounds of thrust, carried the astronauts and their bounty into lunar orbit, where they maneuvered and docked with the Command Module, allowing the three Moon travelers to reunite and return home. The LM was an incredible craft, and in my opinion, the most advanced, sophisticated, and flawless spaceship yet built. It proved itself a true hero during the crisis of Apollo 13. Though I've never worked for the Grumman Corporation, the company who designed and built the Lunar Module, I admire them greatly.

This next mission in the cycle, Apollo 9 with astronauts Jim McDivitt, Dave Scott, and Rusty Schweickart on board and the LM stored below, lifted off from the Kennedy Space Center on March 3, 1969. The crew was charged with fully assessing this new breed of machine in orbit, and it was by far the most difficult assignment to date: test-driving the first manned spaceship designed to travel only in space. Four days later, on March 7, 1969, the LM completed its inaugural flight with flying colors and was pronounced fit to attempt a second test, this time near the Moon. On March 13, McDivitt, Scott, and Schweickart abandoned their LM in orbit, fired the Service Module engine to begin their return to Earth, and separated from the SM, leaving it to burn up in the atmosphere. They rotated the CM flat end forward so that the heatshield would absorb the scorching flames of reentry. The entry guidance computer program worked perfectly, as did our newly revised crew procedures to monitor it, and the Command Module splashed down into the Atlantic Ocean in sight of the waiting recovery ship. The Apollo 9 crew and their mission have received comparatively little attention since that time, yet they were the trailblazers who cleared our path to the Moon.

"You Need a Different Plan"

As I was to discover many times in my career, smugness often summons its own repayment. This particular one came due while I was supporting NASA during a briefing to the crew members of an upcoming flight. Not just any flight, but the big one, Mission G, also known as Apollo 11. From the engineering notes that I kept in those days, the meeting occurred in April 1969, several months before the launch itself, and about a month before the next scheduled flight, Apollo 10, lifted off.

This could have been a refresher briefing for the Apollo 11 crew, bringing them up to date on last-minute adjustments we'd made to their procedures for their own flight, now three months away. Only one of the astronauts from the crew attended the briefing, the Command Module Pilot. Since he was the one who had to fly it, the CM Pilot was charged with acquiring an in-depth understanding of how the onboard computer would handle reentry and what to do about it if it went astray.

So my schooling that day came at the hands of either Michael Collins, CM Pilot of the prime crew (the astronaut who later went to the Moon on that flight), or Ken Mattingly, the CM Pilot of the backup crew. I think it was Mattingly because he'd been a late addition to the Apollo 11 backup crew and was likely focused on catching up. The memory of my comedown that day is sharper now than of the agent who delivered it.

Using my "experience" piloting the Command Module simulator through numerous reentries, I'd developed a specific procedure for the crew to use to ensure that the onboard computer was operating properly during the spacecraft's initial plunge into the atmosphere. I recommended that the Command Module Pilot monitor the DSKY displays for a specific set of data values, and when the spacecraft reached its maximum g-force, 6.5 g's, at about two and a half minutes into reentry, he should use the keyboard to call up a different set of data and compare those to some reference values we'd included in the charts. As I went on describing my recommendation, Ken Mattingly sat quietly, listening and reading our document.

When I finally stopped talking he looked up at me and asked, "Son, you're not a pilot, are you?"

I was caught off guard by his question. I shook my head.

"Let me explain something. At 6 g's I'll be pretty much flattened into my seat. I can read the displays, and if I need to, I can use the control stick, since it's by my leg. But I'll be damned if I'm going to reach up and punch in some numbers on the keyboard."

I waited, not saying anything, and not having much to say.

He looked down at my charts again and said, "I suggest you come up with a different plan." The meeting was over.

I did.

Apollo 10: The Trial Run

The next mission in the sequence was Apollo 10, scheduled for launch in May 1969. By then I was deep into my understanding of how the entry guidance system worked, and how to lay out the complicated crew procedures to follow and actions to take in case of a problem during reentry. Phil Moseley and I even put together the set of charts (modified per my embarrassing briefing of the Apollo 11 CM Pilot) that those three astronauts would carry with them to refer to during that flight. In many ways, Apollo 10 would duplicate much of the work first tried on Apollo 9, but with one big difference; this time they were going to the Moon to do it.

Short of actually touching down on the lunar surface, this flight was to be the final and full-fledged dress rehearsal for an actual Moon landing scheduled for the next mission, and NASA took no chances with the crew. They were all experienced spacemen. The Commander, Tom Stafford, had already flown two NASA missions: Gemini 6 with Wally Schirra in December 1965, and Gemini 9 with Gene Cernan in June 1966. The Command Module Pilot, John Young, had also flown in space twice: Gemini 3 with Gus Grissom in March 1965, and Gemini 10 with Michael Collins, later of Apollo 11 fame, in July 1966. The Lunar Module Pilot was Gene Cernan, who had cut his teeth under the command of Tom Stafford on Gemini 9.

Apollo 10 left for the Moon on May 18, 1969, and performed a nearly

Apollo 10 Crew (*left to right*): Eugene A. Cernan, Thomas P. Stafford, and John W. Young, 1969. NASA Image.

flawless eight-day checkout run for the Moon landing to come. I say "nearly," as at one point the Lunar Module briefly spun out of control when the crew fired its ascent engine with a switch in the wrong position. On May 26, the Command Module and its crew returned to Earth by parachuting into the Pacific. It was the most complex and ambitious human spaceflight ever attempted and was judged a success across the board. Now Apollo 11 was ready to place the first human footprints on the Moon.

Apollo 11: Earning Our Pay

I felt I was contributing. I was up to speed. Phil Moseley and I coauthored the "Recommended Entry Monitoring and Backup Control Procedures for Apollo 11 (Mission G)" document. For all my fussing about NASA-speak, the title speaks for itself. It contained everything that Neil Armstrong, Michael Collins, and Buzz Aldrin needed to know, including the set of "cheat-sheet" charts they'd take with them to the Moon and back, to ensure that during reentry they'd land safely back on Earth. As we discovered later, it turned out that the Apollo 11 crew had to use our new procedures during their reentry and found the work we'd done was solid. It doesn't get better than that.

The weeks leading up to the launch of Apollo 11 were a haze of frenzied work, mounting anxiety, and also a surreal euphoria of expectation. Getting our procedures into the crew's hands was only part of the job. During the flight itself, we could be called in at any hour to support our NASA counterparts who'd be working around-the-clock shifts in back room support at the Manned Spacecraft Center, keeping track of how closely the Apollo spacecraft was adhering to its nominal trajectory during its return to Earth, and supporting the Mission Control flight controller in determining any adjustments needed to keep the crew safely within their entry corridor. Preparing for that meant we had to take advantage of whatever little RTCC time might be available

Apollo 11 Crew (*left to right*): Neil A. Armstrong, Michael Collins, and Edwin E. "Buzz" Aldrin Jr., 1969. NASA Image.

during the weeks prior to launch to make last minute computer runs to bring our own reference charts up to date with NASA's latest launch and return plans for the flight.

When I wasn't working on the preparations for Apollo 11, I was worrying about it. This would be the third human flight to the Moon in seven months, and the first one to try to land. I kept envisioning the misstep on the last one, the Apollo 10 Lunar Module spinning out of control when the crew fired its ascent engine. If that happened on this flight the crew was doomed. Then there was the reentry. While I was delighted to have played a role in working up a completely new set of procedures for that phase, I was also apprehensive. Being responsible for them meant if they were wrong it was because of me. Short of a catastrophic but undetected error in the onboard computer (or if somehow, in spite of everything, our procedures for taking manual control weren't correct), the astronauts' lives would likely not be endangered during reentry if I'd made a mistake. But the prospect of flying an unexpected roller coaster ride through the atmosphere was a possibility, with a consequence that the Command Module splashdown might occur hundreds of miles away from the nearest recovery ship. That could be hazardous to the crew; the Apollo Command Module was not a terribly dependable boat.

And while everyone was putting in a lot of hours at work, there were still the mundane chores to attend to: mowing the lawn, shopping for groceries . . . and meeting with the obstetrician. Several months earlier, Barbara and I had learned she was pregnant with our first child, with the expected delivery late in November. We were excited—"over the moon," so to speak—but we were also moving into unexplored and disquieting territory for us. The parallels in my own life to NASA's landing the first man on the Moon weren't lost on me. The hell of it was, because of what my job required just then, Barbara was left alone to deal with the quotidian stresses a new pregnancy imposes. My role was mostly to worry during the day and catch up with her in the evenings.

As the first week in July rolled around, just getting to work became a challenge. At that time we lived in a rented duplex located in Seabrook, Texas, about six miles from TRW. Seabrook was a small and sleepy town back then, as was the two-lane NASA Road 1 winding between it and my office. That was a fifteen-minute drive most days. But starting around July 4 that year, the bottlenecks sometimes rivaled anything in a major U.S. city. At times it seemed that everyone was watching us. Television, radio, and newspaper reporters from around the world descended on Clear Lake City to cover the story; visitors may have outnumbered residents during those three weeks in July.

Not that we were immune to the excitement. In spite of the around-the-clock workload and the unrelenting anxiety over getting it all right, the fervor of anticipation that was building worldwide was intensified in everyone who worked and lived in Clear Lake City. It was hard to think about anything else. In a few days Walter Cronkite would be narrating, live on national TV, just how well we had done our job. We read the latest Apollo stories in the morning newspaper over breakfast while local TV news anchors gabbed about it in the background. We'd break away from our desks whenever we could to stand in front of the TV set in the employee cafeteria. We switched on our own TV as soon as we got home at night and sat on the sofa eating dinner on trays. Out-of-town family members called to see if we knew any particulars that hadn't made it into the news.

July 16 arrived. Wednesday. Launch day. Liftoff was scheduled for 8:32 a.m. local time. I didn't get much sleep the night before. I woke up early and turned on the TV. No launch scrub had been announced; everything was still "Go." I had little appetite. My breakfast was coffee and donuts. I decided I wasn't going in to work until the launch was complete and the astronauts were on their way. The next two hours felt like days; TV broadcasters supplied hundreds of details, many banal, some just wrong. The final countdown began. T minus sixteen seconds. Over on CBS Walter Cronkite and now retired Apollo 7 Commander Wally Schirra were reporting live from the Kennedy Space Center. Liftoff! The Saturn V rocket, with Neil Armstrong, Mike Collins, and Buzz Aldrin riding atop it, bellowed and lumbered into an impossibly blue Florida sky. "What a moment," Cronkite murmured. What a moment! I watched the rocket ship arc up and over and away and shrink until it disappeared from view. Another ten minutes passed and a NASA voice announced: "Earth orbit insertion." Apollo 11 had made it safely to orbit. It was time to shower, dress, and get in to the office.

Planning for reentry started immediately. NASA called TRW with their latest timeline for trans-earth injection, the specific time six days hence when Apollo 11 would fire the Service Module engine to take them out of lunar orbit to start their journey home, and the resulting time and location of the start of reentry two days after that. We checked that against the nominal time we'd been using to see if our reentry specifics needed to be tweaked. We'd need to do that several more times as the flight progressed. As I said, every reentry roller coaster ride was different.

July 20, the day scheduled for lunar landing, was a Sunday. The undocking of the Lunar Module from the Command Module, the beginning of the long circular descent to the lunar surface, would happen around noon Houston time, and final touchdown at 3:15 p.m. There were a lot of people working that weekend, but I wasn't one of them; my next duty would start when the astronauts got ready to return. So I had planned to stay home and watch it. I was still early in my career and my take-home pay reflected it; we'd outfitted our small duplex unit with inexpensive but serviceable furniture. We had a small color TV, a twelve-inch RCA that we'd bought for a few hundred dollars, perched on a TV stand that was pressed against one wall of our living room. The sofa was pushed against the opposite wall. Moving that television around wasn't an option; it was connected to a rooftop antenna by a flat wire lead that ran up through the wall—this was well before cable or satellite TV: all the programs were broadcast. I pulled the TV stand as close to our sofa as I could, maybe four feet out from the wall, leaving the power cord stretched behind it like a flea circus trapeze wire, so that we could sit down to watch a compressed, grainy, black and white transmission of Neil Armstrong stepping onto the Moon. I moved the TV on Saturday morning, since we'd be glued to it all weekend.

Anyone who knows me will tell you I get clumsy when I hurry, which I do quite a lot; that wasn't a trait I developed with age. For a reason I can no longer recollect, at some point that day I rushed around the back of that TV stand, the power cord caught my shoe as I stepped over it, and the cord snapped off. I remember trying to catch my balance and keep the TV from toppling at the same time. Fortunately I was successful at both, but I now owned a lifeless TV on the most important viewing weekend of my life.

Something is worth mentioning here. In those days, an owner could open the back of a television set and make minor repairs and replace some parts at home, and by that time I'd finally grown more confident and experienced in doing so. It's amazing that being unable to afford a repairman can motivate you to learn to do it yourself (a skill that has since atrophied through disuse). Radio Shack was the place to get those parts, and one of their stores was located a couple of miles from where we lived. I jumped into my car and rushed over there.

I need to describe that car. It was a dark blue 1969 Volvo 142S two-door stick-shift sedan. We'd bought it new, the original price about $3,000 (about $20,000 today). Though my salary was modest, about $7,500 a year, I splurged on the car for one reason—then, as now, Volvo was one of the safest cars on the road.

The trip to Radio Shack was successful; they had the replacement power cord for my TV and tutored me on how to install it. My weekend saved, I paid for the part and left. NASA Road 1 was narrow, heavily treed, two lanes and winding. It was also, however, a major artery for the surrounding towns, and many small businesses and restaurants were tucked in along it on both sides. The speed limit was probably too high for the traffic, about 40 mph as I recall, and I was doing all of that in my rush to get home. I was about halfway there when a large Cadillac sedan emerged from a driveway on my right, hidden by trees, and about a hundred feet in front of me. The driver stopped in my lane, his car broadside to mine. I think he was waiting for traffic coming the other way to clear so that he could turn left. As I caution kids in my talk on "Why Science Matters," a car going at 40 miles per hour is traveling about 60 feet a second, which means that even if it takes you only one second to slam on the brakes, you've already gone that far.

I discovered the physics for myself that day. I hit my brakes and left a stretch of matching tread marks in my wake. The Volvo skidded but didn't spin out. I collided head-on into the driver's side of the Cadillac, and from that point on everything seemed to happen in an unnaturally protracted instant. The front end of my car collapsed, ramming the bumper into the grill, the grill into the radiator, and the radiator into the engine. Both front fenders accordioned, bowing outward and absorbing much of the impact. I wasn't wearing my seat belt so when my car stopped, I didn't. My chest slammed against the steering wheel and it gave way. The round outer rim folded down and away from my arms like a limp pancake, and the center column collapsed through the firewall into the engine compartment, absorbing the blow to my chest. When my Volvo finally came to rest, its engine quiet except for the hiss of escaping steam, I was dazed but, as far as I could tell, unhurt.

Time restarted at its accustomed pace and I tried to clear my head. The Cadillac was shoved sideways several feet from the impact, the driver's window was shattered, and the door panel caved inward. I couldn't see the driver. I tried to get out but my door wouldn't open. I slid across the seat and tried the other one, but it was jammed closed too. I managed to roll down the passenger window and crawled out through it. Once outside I had to grab the edge of the roof to keep from swaying. I don't know if anybody stopped to help out; they probably did. I do remember that a police car showed up, and not long after that, an ambulance. The cop asked me questions, which I tried to answer, while the ambulance crew got the Cadillac's door open and lifted the driver onto a rolling stretcher: an older man, probably in his seventies from the lines of his face and the white of his hair, but he appeared to be conscious. The policeman went over to talk to a passenger in the car, who I think was a woman, maybe the driver's wife, and then the ambulance sped away with its lights flashing. The cop came back, walked past my Volvo and stepped off the lengths of my skid marks. He came to talk to me some more. Two tow trucks showed up. I don't remember where I got a pencil and paper, but someone wanted to see my insurance card and asked for my address and phone number—that might have been the tow driver assigned to me, but my hands were shaking so badly I couldn't hold the pencil steady to write. My Volvo was winched up onto the truck, the entire front section bent and shattered, and towed away. Despite my shock, I was thinking clearly enough to retrieve the Radio Shack bag with my replacement power cord in it.

The policeman finished his investigation and asked if he could give me a ride home. I hadn't considered how I was going to get there otherwise. This was well before cell phones, and anyway, we didn't own a second car, so I was grateful for his offer. On the drive home, he asked me if I'd been speeding. I told him I didn't think so. He said the length of my tread marks showed I might have been, but in any case, the driver of the Cadillac was at fault for having pulled out into traffic and stopped, so he wasn't going to cite me. I was relieved; this was before the days of no-fault insurance, and a speeding ticket would complicate any insurance payment I'd collect to replace my car. He dropped me off in front of our duplex. Barbara was waiting for me on the porch; she'd been alarmed when I hadn't returned from my errand and was even more so when the cop car pulled up outside. We went inside and I tried to tell her what had happened, but even then I was still foggy. I was sore, I had some bruises, mostly on my arms, and my body ached for several days afterward. But I was all right. I'd been through a horrific accident, the other guy was taken away in an ambulance, and yet I was unhurt. That Volvo had behaved exactly as it had been designed to do in an accident, and it rescued me from serious injury and maybe even death.

I've often thought about what might have happened had Barbara decided to go with me that day. The collapsing steering wheel kept me from colliding with the dashboard and windshield. She and our unborn child wouldn't have had that protection. In addition to all of its other safety features, Volvo had been the earliest car manufacturer to provide three-point seat belts as standard equipment. But wearing one wasn't required until 1983, so I didn't do so that day, and that car protected me in spite of myself. I've worn seat belts religiously ever since. I never found out how the man driving the Cadillac had fared; I hold on to the memory of seeing him conscious when the ambulance took him off.

In spite of the extensive damage to my car, including destruction of the engine, warping of the body, and severe bending of the frame, later on neither his insurance company nor mine would pay enough to replace my Volvo. I got it repaired as best it could be and rented a loaner car in the interim to use during the three months those repairs required. The Volvo came back to me looking much better, and still serviceable, but it was never the same. Soon after, some of the engine

My son Paul helping me wash our Volvo 142S, 1970.

parts failed, and later on so did the transmission. The car was in constant need of servicing, so a few years later I traded it in. Having to do that was an emotional moment; I was parting with an aging, damaged friend that had saved my life.

Later in the afternoon of the day of the accident, after I had had time to calm down, I replaced the power cord on the TV and we got our news coverage back. Then on Sunday, July 20, at 9:56 p.m., we watched the live broadcast as Neil Armstrong stepped out onto the Moon.

Mankind's first excursion on another world was a relatively brief one. Armstrong and Aldrin collected soil samples, inspected the LM, set up three experiments, took numerous photos, erected an American flag, received congratulations from President Richard Nixon, ascended the ladder, and crawled back inside the Lunar Module, all completed in two hours and thirty minutes, about the time it takes to watch one of the *Star Wars* films. But by a little past midnight Houston time on

Monday, July 21, 1969, the world was forever changed. This was a storyline of space adventure like no other in history, before or since.

The Apollo 11 astronauts spent an additional thirteen hours on the Moon sealed inside the lander's cabin, resting, eating, reflecting, and preparing to leave. On Monday, July 21, at 12:54 p.m. Houston time, the LM ascent engine fired, lifting Apollo 11 off the descent stage and away from the Moon. At 6:00 p.m. the Lunar Module ascent stage docked with the Command and Service Module, and Armstrong and Aldrin rejoined Michael Collins inside. The ascent stage was jettisoned and abandoned to drift freely in space. Its final resting place is unknown. As midnight approached on that special Monday in Houston, the CSM fired its engine for two and a half minutes and Apollo 11 left the orbit of the Moon and headed home. It was time to prepare for reentry. I needed to get to work.

On Tuesday morning I was requested by NASA, as some others of the TRW team had been, to report to one of the small support rooms that encircled the Manned Spacecraft Center's Mission Control Room. Our NASA Task Monitor, Dave Heath, who was responsible for ensuring Apollo 11's reentry, would be working shifts in there until splashdown on Thursday. For this momentous flight, a few support contractors like me were asked to be in the room to help. No mission ever followed its

Neil Armstrong inside the Apollo 11 Lunar Module after his moonwalk, July 20, 1969. NASA Image.

planned return trajectory precisely, and that included Apollo 11. Small variations in the amount of time spent in lunar orbit, or an altered firing time, or duration of the trans-earth injection burn—all within the bounds of "nominal" flight—would send the Apollo crew homeward on a different trajectory than the one we had calculated, and might even influence where and how reentry occurred.

By analogy, if you're out at night with a flashlight and direct its beam at a darkened roadside billboard, a small flick of your wrist will send the beam of light sweeping across the sign. So, too, could minor flight differences cause alterations in the trajectory that might make the reentry angle into Earth's atmosphere too shallow, or too steep, or the Command Module splashdown too far removed from the waiting recovery ship.

As new tracking information came in from the spacecraft and ground stations, our job was to calculate the effects on entry. If warranted, the astronauts could fire the spacecraft engine one or more times to reposition its homeward trajectory a bit. These were called midcourse corrections, and at 3:00 p.m. on Tuesday, July 22, Apollo 11 was requested to execute a short one, just ten seconds, to bring it back in line. They did so, and our new calculations showed they were now right on target. That's what our NASA team was there for.

Wednesday, July 23, was a relatively quiet day for the reentry folks. No additional correction burns were required, but some nasty weather had cropped up in the Pacific Ocean near where the Apollo 11 splashdown was planned. After some back and forth discussion, NASA decided to move the splashdown point 215 miles farther down-range where a recovery ship, the aircraft carrier USS *Hornet*, would be repositioned. The weather there was expected to be clear. It was too late then for another midcourse correction burn, as the latest that could occur was thirty hours before reentry, so the Apollo Guidance Computer would have to do its special brand of flying to get there.

Flying the reentry that additional distance, 1,500 nautical miles versus the nominal 1,285 nautical miles, was well within the capability of the Command Module's onboard guidance system. We had simulated that particular roller coaster ride repeatedly, and the Apollo crew had trained for it. But it was significantly dissimilar from the one originally planned and required different procedures for the crew to use to ensure that the guidance was working properly, procedures that Phil Moseley and I had recently revised and were responsible for. I kept thinking about NASA's displeasure with the earlier set TRW had produced.

When I woke up that morning, however, that wasn't first on my mind. The evening before I had gotten a call from our company's public relations person requesting me—ordering me, actually—to give an interview to NHK, the Japanese national television network, at 7:00 the next morning. I was surprised that it had come to me, a low-level twenty-five-year-old engineer. I discovered that the NASA brass, and my TRW manager Bob Manders, and his manager Bob Morris had already been corralled by the major U.S. and international broadcasters. On this particular morning I was the last man standing, so to speak.

NBC had set up an elaborate broadcast booth on the roof of the Nassau Bay Hotel, across the street from the Manned Spacecraft Center. Because of the early hour of the interview, they were letting NHK borrow the booth. I pulled together a couple of charts, pasted a cut-out image of Earth onto a sheet of stock paper, made a crude diagram of the Command Module intersecting its atmosphere, and set my alarm. The next morning I got dressed in my only suit and off I went. I arrived on the NBC set, with its wall-sized windows overlooking the NASA Manned Spacecraft Center, familiar to me from many earlier broadcasts, and found it had been completely made over with NHK logos and personnel. Tokyo is fourteen hours ahead of Houston, and I was being interviewed for a live evening television broadcast in Japan. The Japanese broadcast manager welcomed me, I sat down, and the broadcast went live. The moderator spoke to the camera in Japanese, introducing me I think, and then turned to me and asked a question, in perfect English, about how an Apollo reentry worked.

I held up my charts. I described the Command Module as shaped like a Hershey's Kiss. I gabbled on a bit about doing a belly flop in a swimming pool. The moderator nodded and smiled, and turned to the camera and translated what I'd said (though he seemed to use far fewer words than I had). He asked me another question, and this time I showed him how reentry steering worked, using my example of holding a flattened palm into the wind outside a car window. We went on like that, back and forth, for about five minutes, and the interview was at an end.

The NHK team seemed pleased enough with my performance. Afterward, as a token of thanks, they presented me with a small white handkerchief with NHK embroidered on it in orange. I still have it. But I wonder what they made of my amateurish-looking charts and my technical gobbledygook—I couldn't even explain that to my mother in those days. How well did all that translate into Japanese? Had I left TV viewers in Japan thinking the Apollo Command Module's heatshield was made of chocolate?

After the interview I went home, changed clothes, and drove over to the Mission Control Center. I was anxious to get back to the support room. We were still working on the revised reentry information that would be relayed to the Apollo 11 crew to account for the longer flight to splashdown. The data would be voiced up to Mike Collins, the Command Module Pilot, at 8:30 the next morning, three hours before reentry began. There were thirty updated numbers that Collins needed in order to fly that 1,500-nautical-mile reentry and confirm the guidance system's behavior as he did so. We spent the rest of the day checking and rechecking and checking them again to be certain we got them right.

But NHK wasn't done with me yet. At the end of my interview that morning they'd asked if I'd do another one for a radio broadcast that afternoon. We picked a time to meet in the NASA's visitor center, located elsewhere on the grounds, and when the appointed hour came I excused myself to the support room folks and walked across the campus to meet with the NHK radio news reporter, someone different from the people I'd met that morning. We had a question and answer session that lasted about half an hour, including pauses for the reporter to translate, and I went through much the same information that I'd given to their TV host that morning, but without the visual aids. The reporter thanked me profusely and seemed genuinely pleased, though there was no handkerchief offered this time. However confusing it might have been in Japanese, I had confirmation later that my explanation of reentry was correct.

The following morning, Thursday, July 24, 1969, at 11:51 a.m. Houston time, Apollo 11 completed a safe reentry and splashed down in the Pacific Ocean, thirteen nautical miles from the USS *Hornet*. A Navy helicopter with frogmen was on the scene, and a little more than thirty minutes later, astronauts Armstrong, Aldrin, and Collins were on board the helicopter on their way to the recovery ship. In spite of the late shift of the splashdown location by 215 miles, the Apollo onboard guidance system handled the change perfectly.

We got feedback later from Mike Collins that he had used our entry monitoring procedures to assess the computer's behavior during the altered reentry trajectory. Neil Armstrong had expressed reservations to him about trusting them that day. Having trained using the early ones that TRW had supplied, he thought Collins should take control away from the autopilot and fly the entire reentry manually. In fact, those old procedures for handling Apollo 11's additional 215 miles would have instructed Collins to take control of the flight, which meant he'd have had to use less precise indicators to navigate to his splashdown target.

But Collins had trained with our revised version and was comfortable using its procedures and letting the Apollo guidance computer handle the job. Since he was the Command Module Pilot, the one most skilled in the workings of the Command Module's onboard guidance system, his judgment prevailed. Our new and improved procedures told him that the guidance system would do just fine. It did, and the Command Module dropped down as planned near the Navy ship. I felt I'd earned my pay that day!

Apollo 12: "SCE to AUX"

I couldn't spend time savoring the success, since the next scheduled Moon launch was less than four months away. Although the Apollo 11 flight had been scheduled for July, NASA didn't want to risk missing JFK's challenge to put a man on the Moon by the end of the decade if something prevented that first crew from Moon walking. So a second attempt, Apollo 12, was scheduled for launch on November 14, 1969.

After the success of the first Moon landing, NASA's funding to us shrank a bit, so some of our colleagues at TRW moved on to other

Apollo 12 Crew (*left to right*): Charles "Pete" Conrad Jr., Richard F. Gordon, and Alan L. Bean, 1969. NASA Image.

projects or other companies. One of those was an engineer who had been doing reentry planning, so the rest of us were stretched and on a tight deadline; we had to prepare for two upcoming flights and finish the post-flight analyses for Apollo 11. I'd been assigned as responsible engineer for Apollo 12, which meant that all the reentry planning for that flight was on me.

In those days TRW was required to document all of our Apollo work formally and deliver it to NASA. Between August and November 1969 I wrote ten separate documents for Apollo 12, covering everything from crew training to revised entry monitoring procedures, descriptions and analyses of some of the specific, if esoteric, details of flying and controlling that mission's reentry flight path. I've kept the records and copies of some of the documents, which is why I can reconstruct what I was doing back then. There was a lot of work to be done in preparing for an Apollo flight, and I was only one of perhaps fifty engineers at our company, and of four hundred thousand people in total, laboring away at it.

On Friday, November 14, at 10:21 a.m. Houston time, Apollo 12 lifted off from the Kennedy Space Center in Florida with Pete Conrad, Dick

Gordon, and Alan Bean aboard. Five days later, when Apollo 12 Commander Pete Conrad stepped out onto the lunar surface, his first words were, "Whoopee! Man, that may have been a small one for Neil, but that's a long one for me." Though he was joking about his height (Conrad was five inches shorter than Neil Armstrong), he might equally have been talking about his flight. The second lunar landing mission nearly ended seconds after it was launched.

At the beginning of this account, I described the heroics of NASA Flight Controller John Aaron in saving both the crew and the flight of Apollo 12. When lightning struck the Saturn V rocket carrying Apollo 12 into orbit, it shut down all transmissions except voice between the spacecraft and Mission Control in Houston, and it left the Command Module itself running on emergency batteries with a limited duration. John's clear-headed thinking in this demanding situation not only prevented what might have been a hazardous flight abort but also allowed Apollo 12 to continue its mission to the Moon.

John's responsibility included the electrical, environmental, and communications systems on the spacecraft, a controller position known as EECOM. He was stationed at one of twenty flight controller

NASA Flight Controller John Aaron at the Mission Control Room EECOM position. NASA Image.

positions, each responsible for the performance of a specific subsystem during a spaceflight. While the rest of the flight controller team expected John to call for a launch abort—his responsibility in the case of a total electrical failure—he instead requested the crew to reposition a little-used cockpit switch, the Signal Conditioning Electronics (SCE), to its Auxiliary (AUX) position.

Though no one else in Mission Control, nor the astronauts, knew what that switch did, and the astronauts first had to locate it and then trust that changing its setting would make a difference, they followed John's instruction and repositioned the switch. That reinstated the transmission of correct readouts from the onboard electrical systems, opening a window that allowed John to see the real problem. The spacecraft's source of power, the fuel cells, had somehow been knocked offline (the Mission Control team didn't find out about the lightning strike until later). John recommended that the crew reset the fuel cells; that is, turn them back on. All the power and complete data transmission were restored, and Apollo 12 was able to continue with its mission to land on the Moon.

Because of his specific Mission Control responsibility, John alone understood what the SCE switch did and figured out in those few seconds that it could possibly restore valid data so that he could understand how to recover from the power problem onboard. John had noticed the odd behavior of the Signal Conditioning Electronics during a test in an Apollo Command Module at the Kennedy Space Center nearly a year earlier, when that spacecraft was inadvertently placed in the same power "brownout" condition as the one that resulted from the Apollo 12 lightning strike, and that idiosyncrasy had stuck in his mind.

I've often told others that story of how John knew what to do, but here is how John told it to me himself.

> A year earlier, during a live test of the Apollo 7 spacecraft, the ground crew at the Kennedy Space Center had inadvertently switched all of the spacecraft's power load onto one low-power reentry battery. That caused a low voltage "brownout" in the Apollo 7 spacecraft.
>
> Simultaneously, I saw a nonsensical pattern of data values frozen on my screen in Mission Control in Houston, where I was monitoring

> the ground test. My curiosity got the best of me and I spent the next day back in the office understanding what I had witnessed, determining it was due to the Signal Conditioning Electronics equipment doing a low voltage shutdown. The SCE contained a low voltage protection that would cause an automatic SCE shutdown if voltage fell below a certain value. Selecting AUX on the switch bypassed that feature and applied power directly to the internal electronics at whatever voltage was available. Interesting that the readouts on Apollo 12 froze at the exact same values, and with the same pattern, that I remembered from the ground test a year before.

John's colleagues in Mission Control were startled when he said, "Tell the crew to set the SCE to AUX," yet the training of the team and discipline of these professionals led them to trust the teammate in charge of this decision, and they did.

To put John's quick action in a sharper light, just fifty-nine seconds had elapsed from when the Apollo 12 brownout occurred until John made his "SCE to AUX" call. He had only another ninety seconds left before he would have to abort the mission and blow up the Saturn V rocket. There had never been a launch abort during the Apollo Program, and though there was an option to separate the Command Module carrying the crew and parachute them into the nearby Atlantic, it was extremely risky and might have seriously or fatally injured the astronauts.

In Astronaut Jim Lovell's memoir, *Apollo 13*, he reports that because of John's swift analysis and decision making under incredible pressure during Apollo 12, his teammates came to nickname him "steely-eyed missile man." John Aaron not only saved that flight to the Moon; he saved all five of the Apollo missions to follow. The lunar landing prior to this one, Apollo 11, made Neil Armstrong the first human to walk on the Moon. And when that spacecraft splashed down in the Pacific on July 24, 1969, John F. Kennedy's 1961 challenge to the nation of "the goal, before this decade is out, of landing a man on the Moon and returning him safely to the Earth" had been fulfilled.

Public interest in the program dropped precipitously, and the TV coverage went along with it. Just three months after Apollo 12 splashed down, Congress made severe cuts in funding to the Apollo Program. Had Apollo 12 been aborted at liftoff, the remaining missions would unquestionably have been canceled, and the flights of Apollo 13, 14, 15, 16, and 17 would be only a lost dream. "Steely-eyed missile man" indeed.

John Aaron was central to achieving this nation's voyage of humans into space, and he is also one of my career heroes. I met him later on, in 1981, in his job as NASA's Chief of the Space Software Division at the Johnson Space Center in Houston, when both he and I had moved on from Apollo and were working to help make the Space Shuttle fly. I'll reintroduce you to John later, but I have a story about him to tell now.

A few years ago John told me how it was that he came to work on the space program. His family owned a farm in southwestern Oklahoma, and after finishing high school John wanted to expand it into a cattle ranch. Realizing he would also need a steady income, he enrolled in college to get a science and math degree so that he could become a science teacher. During his senior year in college he found out that graduation wouldn't earn him enough credits for a teaching certificate, and since by then he was married, he needed a job. Someone told him NASA was hiring, so he applied and got an offer to come to Houston. He decided to take it.

"Funny thing is," he told me, "I still thought I would go back home soon and pursue ranching, which I never did, of course. Looking back, it was the best decision that I ever made and one that was forced on me to make."

After that harrowing encounter with lightning, the Apollo 12 spacecraft recovered its bearings and carried our fourth set of lunar space travelers to the Moon. Conrad and Bean put their Lunar Module down in an area called the Ocean of Storms, less than two hundred yards from their intended objective, the unmanned U.S. space probe Surveyor 3, which had been sent there in 1967. During their later excursions, Bean and Conrad retrieved portions of the lifeless probe and returned them to Earth for scientific examination. They stayed on the Moon for more than thirty hours and took two separate forays onto the surface, the first lasting almost four hours and the second well over three and a half. They deployed scientific experiments, acquired core samples, and took numerous photographs.

Apollo 12 rocketed away from the lunar surface on Thursday, November 20, at 8:25 a.m. Houston time. After rendezvousing with the CSM and jettisoning the LM Ascent Stage, they stayed in lunar orbit for another day, capturing photographic records of the surface and other planned landing sites. Then they fired the CSM engine for a little over two minutes and headed home. At that point, the Command and Service Module had been in lunar orbit for eighty-nine hours, more than three and a half days, quite a difference from the more cautious approach dictated for Apollo 11.

Two short midcourse correction burns were required on the return flight to keep Apollo 12 safely within the reentry corridor, the first one of five seconds on Saturday, November 22, 1969, when they were fifty-six hours from home (and also the sixth anniversary of John F. Kennedy's assassination), and another of six seconds just three hours before the start of reentry on Monday, November 24. The astronauts' return trip was adequately "nominal" so that I could support Dave Heath in the NASA Mission Control back room from my desk at TRW. He would call me over there if he needed me, and for this flight, he didn't. Reentry began at 2:45 p.m. Houston time, and fourteen minutes later Apollo 12 landed safely in the Pacific Ocean just two and a half miles from the recovery ship, close enough for the Navy crew of the USS *Hornet* to photograph the all-Navy team of astronauts splashing down.

During the first several Apollo missions, there was concern at NASA that the astronauts might pick up some alien bug while on the Moon's surface. So when they returned to Earth and were delivered onto the aircraft carrier, they were immediately placed in quarantine in a structure called the Mobile Quarantine Facility (MQF) lest they unleash some contagion onto a non-immune populace.

The quarantine facility looked very much like a classic Airstream trailer, which it was. It was nine feet wide and thirty-five feet long with a living area, sleeping section, and kitchen. It had been modified to keep its internal pressure lower than that outside, to make sure that any germs got sucked in, not out, and it had a specialized air filtration system. It was also structurally reinforced so that it could be transported by the aircraft carrier to Hawaii, then offloaded onto an Air Force cargo plane and flown, with the astronauts still inside, to Ellington Air

Force Base in Houston. Once there, it was hauled down the road to the Manned Spacecraft Center, a journey that took five days in total from Pacific Ocean splashdown to Houston.

When the quarantined spacemen were back home, there was more to come. The crew was transferred into a long-term quarantine building, called the Lunar Receiving Laboratory (LRL), where they remained isolated for an additional ten days, or until they were given an "all clear" by the doctors and scientists in attendance. All the equipment that was used on the lunar surface and returned, and all the moon rocks that the astronauts brought home, were put in there too. The returning heroes' post-flight celebrating was done while they were locked inside a trailer. To the relief of the entire astronaut corps, this practice was discontinued after Apollo 14.

During those sixteen days in 1969, from November 24 when Apollo 12 splashed down to December 10, when Conrad, Bean, and Gordon were released from confinement, I was celebrating too, and for once I was absorbed with my personal life and not with NASA events. On November 29 my first son, Paul, was born, at nearly the same time as the Apollo 12 crew and their MQF landed at Ellington. His birth was the beginning of a new journey of discovery for me, and one from which I've never stopped learning.

1970: The Ticking Clock Slows

Preparing for these early Apollo missions was exhausting. Beginning with Apollo 7, NASA launched an Apollo every two months until Apollo 11 landed on the Moon in July 1969. That schedule meant three or four astronaut crews were training for their flights at the same time. Since every Apollo mission was unique, reentry procedures had to be developed and tested for each one. Apollo 8, for example, coming back from the Moon, reentered Earth's atmosphere at about 24,600 mph, while the next mission, Apollo 9, confined to Earth orbit, reentered at about 17,500 mph. These speeds and resulting temperature buildup on the heatshields placed differing demands on the entry guidance computer and produced very different rides for their crews to monitor. Apollo 8 returned to Earth in December when the atmosphere was colder and denser in the Northern Hemisphere, and Apollo 10 returned in May when it was less so. That difference in density, or thickness, of the air affected how much drag and heating the Command Module experienced during reentry, again changing the roller coaster ride.

Our reentry planning team was kept busy juggling a number of balls of assorted sizes and shapes at the same time, and we found ourselves riding our own sort of roller coaster. The weeks and days leading up to a launch were extremely intense, and we were sometimes called upon to support NASA during the mission itself if something off-nominal arose. So when a mission finally ended with a successful splashdown, we were ready to go out to celebrate with pizza and beer. Unfortunately, that adrenaline rush was quickly followed by a plunge of intensity that often lingered well past a weekend.

Because NASA had built the Manned Spacecraft Center out in the pastures, everyone and everything required by the people who worked there had sprung up around it. All of us—NASA folks, astronauts, and contractors alike—lived in the same neighborhoods, shopped at the same Randall's grocery store, went to the same churches, cheered for our kids in the same All-Play league. The astronauts played as hard as they worked. After every successful mission, astronauts and employees of NASA and the contractors would congregate at local watering holes for splashdown parties, a ritual adopted from the early days of launch celebrations at Cape Canaveral. Most of these were closed to the public; to be admitted you had to prove you worked on the program. In the sequestered community surrounding the Space Center, that meant pretty much anyone except a reporter, which was good, since these parties often grew boisterous as the evening wore on and more celebratory spirits splashed down. They were a community and family celebration of success.

Then, following Apollo 11's successful landing on the Moon and the fulfillment of JFK's challenge, NASA eased up on the launch schedule a bit. Apollo 12 launched four months later, although still in 1969. That mission had always been a second chance to get to the Moon by the end of the decade. With that "ticking clock" urgency in NASA relieved, the next five launches, Apollo 13 through Apollo 17, were stretched out over the following three years.

After I'd been through a couple of these, I slipped into a "procrastinate and then shift to crises-mode" behavior. Even though another launch might be only months away, and every mission had prodigious planning work required, I couldn't get myself into gear to do it. Sometimes my listlessness went on for several weeks. I'd make an appearance of laboring away at my desk every day, but I had little to show for it, and I knew it. It wasn't as though the folks I worked with were pressing me for results either; everybody was pretty much burned out by then. But as the next due date closed in, my dormant anxiety would arise and I'd shift into crash mode, realizing now that I had to work nights and weekends to catch up. Somehow I always did, which only reinforced my bad behavior following the next splashdown. It was a hell of a way to plan a reentry.

After Apollo 11, after Armstrong, Aldrin, and Collins splashed down, the upbeat mood of the TRW team began to wane. I had other parties to attend, ones arranged to say goodbye to coworkers who were leaving. In fact, TRW was forced to hand out a round of layoff notices while Apollo 11 was still traveling back from the Moon. Public interest and congressional funding and support for the Apollo Program plunged once we accomplished President Kennedy's goal.

TRW's contract with NASA had originally called for about eighty technical professionals and engineers. We had teams assigned to every phase of an Apollo mission, from liftoff to landing on the Moon, the return to Earth, reentry, and splashdown. Just like the people on my team, each of the others had worked hand-in-hand with their NASA counterparts to provide plans for that mission phase and ways for the astronauts to recover if something went wrong. But by the time Apollo 11 lifted off and made it into near-Earth orbit, that front part of each flight had become pretty routine; we'd been doing manned launches since the Mercury Program. Not a lot of extra planning was required there anymore, and facing budget cuts themselves, NASA passed on some of the pain to us. More was yet to come; in the months ahead I'd go to farewell parties nearly every week.

Even the second lunar flight, Apollo 12—which accomplished much more than Apollo 11 in terms of the science, technology, and the ability of humans to operate on the Moon—had attracted substantially less press coverage than the first one. A sort-of "oh well, been there, done that" attitude had settled over the American public. The enormous expense of the Apollo Program had been sold as a race against the Soviets for technical and scientific supremacy. We had won the race. Game over. Mission accomplished. The nation was preoccupied with other problems now: the war in Vietnam, escalating inflation, political turmoil, and social upheaval, particularly among young people.

In February 1970 barely seven months after Neil Armstrong's one small step, NASA lost funding for Apollo 20, the last scheduled journey to the Moon, and canceled that flight. In September that year the agency succumbed to additional cuts and canceled the eighteenth and nineteenth planned missions. These spacecraft had been built; the astronauts were available and in training; the missions were ready to go. One estimate of the total savings realized by American taxpayers for canceling these two additional opportunities to explore the Moon was about $42 million in 1970 dollars. Since the overall cost for the six successful Apollo moon landings was $25 billion, this level of savings seemed a modest extra price to pay for the singular opportunity to revisit the Moon. Much of the spacecraft hardware, as well as the astronauts and ground support people, went on to Apollo Skylab missions that followed.

On December 13, 1972, Gene Cernan, the commander of Apollo 17, climbed up the ladder and into his Lunar Module, becoming the last man to walk on the Moon. That was the first time in all of human history that humans could leave the bounds of our planet to walk on another one, and that someone could ask the question, "Should we go back to the Moon?" and have it make sense. Our answer then was "Why bother?" and for many people it is the same today.

Apollo 13: No Hand of God

Though the nation and the news media grew bored with manned spaceflight after Apollo 11, for one week at least, in April 1970, the world briefly returned its attention to Apollo. Apollo 13 was launched from the Kennedy Space Center on Saturday, April 11, 1970. Though Neil Armstrong had walked on the Moon the previous July, this mission was now considered sufficiently routine to attract little press coverage.

Apollo 13 Crew (*left to right*): James A. Lovell Jr., John L. Swigert, and Fred W. Haise Jr., 1970. NASA Image.

Apollo 13 started out well. It lifted off from KSC into a blue Florida sky at 1:13 p.m. Houston time, with astronauts Jim Lovell, Jack Swigert, and Fred Haise on board. I spent that weekend at home. There wasn't much reentry support needed until after the astronauts finished up on the Moon and began the process of returning. My son Paul was just over four months old, and I was trying to focus more of my time on figuring out what it meant to be a husband and a father.

I did have more office work to do than on earlier flights, though. Beginning with Apollo 13, my supervisor Bob Manders named me Task Manager for TRW's reentry planning. The TRW contract with NASA was broken into subunits called tasks, each one focused on a specific element of our contract responsibilities. This one, identified as Task A-220, covered all of the Apollo reentry mission planning and support to which we were assigned. The task involved a few people, two or three engineers and a couple of technical support professionals, plus a "task manager" work supervisor who was supposed to ensure that everything got done correctly and on time.

Each of our tasks had a corresponding NASA Task Monitor who resided at the Manned Spacecraft Center. Fortunately, my task continued to report to Dave Heath, whose direction (or mentoring) had helped me navigate the sea of NASA's expectations early on. These NASA supervisors were far more than government bean counters; they were usually experienced engineers employed by NASA to perform all the planning and backroom support for every Apollo mission. They were the ones who were stationed in the flight support rooms backstage at Mission Control during a flight. The flight controllers in the Mission Control Room had direct access to them twenty-four hours a day while a mission was in progress, and whenever called upon they had to be ready with rapid, detailed, and correct technical information in their fields of specialty. As the support contractor for the NASA Landing Analysis Branch, TRW was an extension of their team; we were expected to add our knowledge and perspective to their duties.

Dave Heath's personal journey to become NASA's reentry specialist for planning for Apollo missions started in 1964. He joined NASA in

My NASA Task Monitor Dave Heath. Image courtesy of Dave Heath.

Houston as a new college graduate while the two-man spaceflights of the Gemini Program—the warm-up for Apollo—were under way. As Dave was leaving for NASA, his physics professor had confided that his own PhD thesis had proved it was impossible to orbit an object around the Earth, but he'd neglected to consider that the rocket launching the object could be built with detachable fuel sections that could be discarded when empty, greatly reducing weight, which then made getting into orbit achievable. "It was a good message," Dave remembered. "Check your assumptions, think outside the box."

Once on the job, Dave was assigned to the team planning the Apollo 1 reentry. He discovered that beyond some classified work being done by the U.S. Air Force, the science and engineering of managing a reentering spacecraft, particularly one shaped like the Apollo Command Module, was in its infancy. Dave took on the task to educate himself on what was known and then to improve on that knowledge. He was then named NASA's Reentry Project Manager for the upcoming Apollo 1 flight, and he became the single point of contact between NASA flight controllers and astronauts and the contractors who were building the Command Module. Eventually NASA assigned Dave the lead reentry responsibility for the all of the Apollo flights.

My job had expanded substantially for Apollo 13, with my own work and a small handful of other people's efforts to shepherd, though we had had almost five months to get everything done. My shepherding job was pretty easy—the individuals assigned to the reentry task were terrific. I didn't need to look over their shoulders or catch their mistakes, as they were a highly experienced, self-disciplined bunch who were as professionally and personally invested as anyone on the planet in making Apollo a success. So beyond getting my own engineering work done, the bulk of my time was taken up working hand-in-hand with Dave Heath and his boss, Jon Harpold.

Dave was under significant pressure from his upper management to complete all his responsibilities on time and to ensure that they meshed perfectly with those of his peers working on other elements of the flight, such as lunar orbit planning and support. Furthermore, his experiences with TRW's performance on this task early on had been a rocky ride. I've mentioned that Neil Armstrong among others had found the original TRW reentry monitoring plan wanting, and that judgment had come down on Dave first, not on us. Evidently his subsequent attempts to work at a solution with my predecessor had not gone well; Dave had given the TRW reentry task a series of failing grades during the time leading up to the Apollo 11 splashdown. So in spite of our work on Apollo 11 and Apollo 12, he wasn't quite ready to declare victory. Part of my new job was to convince him otherwise, and some of that depended on how well we supported him and the new astronaut crew during this upcoming flight.

Apollo 13 soared into Earth orbit precisely on plan, and stayed there for a couple of orbits while the ground team and crew checked that all of their systems were "go" to make a second engine burn that would launch them to the Moon. They did a short TV broadcast during that time, which few people except NASA folks bothered to watch. When Mission Control was satisfied, the crew fired the Saturn V upper stage engine for six minutes to achieve translunar injection, which boosted them away from Earth into an elongated orbit that carried them drifting toward the Moon. The Apollo 13 crew settled in for their three-day coast. They did some housekeeping, took photos of Earth weather patterns, and finally got some sleep at 2:00 a.m. Houston time. I was in bed long before that. Paul had started sleeping through the night, which meant we might be able to as well. It turned out that everybody slept in.

The Apollo 13 crew came back online around noon the next day, Sunday, April 12. The day was to be a relatively quiet one. Another TV broadcast, a midcourse correction burn, and taking photographs of Comet Bennett as it streaked inward through the solar system. The big news story of the day was that the Beatles were breaking up. Capsule Communicator Astronaut Joe Kerwin (the "CapCom," the only person in Mission Control authorized to communicate directly with an Apollo crew) reported to Apollo 13 that the Fab Four had made a half-billion dollars in their short career but "rumors that they will use this money to start their own space program are false." To which Jim Lovell responded, "Maybe we could borrow some." The irony of NASA's funding cutbacks wasn't lost on him.

At 7:30 that night, Lovell, Swigert, and Haise gave the world an unprecedented view of their spacecraft, of the Moon, and of the Earth, via a fifty-minute live television show from space. Jim Lovell demonstrated the strangeness and beauty of zero gravity by pointing his camera at a spray of frozen water droplets outside the spacecraft as they spread away from a waste water dump. "It lights up the whole sky around the Moon," he reported. They broadcast and narrated the midcourse correction burn, which was performed using the CSM engine. Not imagining what was to come, Fred Haise said, "I guess we'll see how about the only system we haven't used yet works [meaning the Lunar Module]. Everything else sure has worked mighty fine."

The four-second midcourse correction went exactly as planned, though it complicated things later on. This engine firing was to take the Apollo 13 spacecraft out of a so-called free return trajectory and into one that would ultimately put them into an orbit around the Moon. The logic behind using a free return trajectory at the outset of a flight was sound. It was an orbit designed to bring the astronauts home if one of their major systems failed. A free-return trajectory was just that, a single engine burn that pushed the spacecraft into a very elongated orbit that extended all the way out to and around the Moon and then circled back again to Earth with no additional engine firings; that is, "free."

A simple earthbound example of a free-return trajectory is tossing a ball in the air. One throw by you will get it started, and no more effort is required to get it back. When the ball gets as high as it's going, gravity takes over and pulls it back down. The ball's return trip to you is "free".

But Apollo 13 was intended to be a Moon landing mission, and since no serious system problems had arisen thus far, the crew fired the big Service Module engine to set them on a course that removed the possibility for a "free return" and that would take them close enough to orbit the Moon itself. At 8:15 p.m. Houston time Apollo 13 ended the live TV show, and the crew went back to some routine tasks. They started their sleep time around 2:30 a.m. In one of many bad breaks for the Apollo 13 crew, just one day later they would have to get back onto a free-return trajectory if they were to be saved.

It was now Monday, April 13. I got up around 7:00 and went in to work by 8:30. We were anticipating a pretty light day. Apollo 13 was still more than a day away from the Moon, and the return trip wouldn't start for another three days after that. The planning for Apollo 14, scheduled for January 1971, was already under way, so we wouldn't be sitting on our hands. We took a break during the day to watch TV and keep up with the doings on Apollo 13, but there was little on except soap operas. I called Dave Heath a couple of times to find out how the flight was progressing.

The Apollo 13 crew expected another light day too, and until very late that day, it was. They woke up around 11:00 a.m. Houston time and did some routine checkout of their systems, including powering up the Lunar Module, and the Commander and Lunar Module Pilot worked their way into it to do a thorough inspection and run more tests. A little after 8:30 that evening, and while they were still inside the LM, the crew held another live broadcast that gave the people of Earth an unprecedented live color television tour of the world's only manned lunar spacecraft.

In motion picture parlance, they provided a long hand-held tracking shot, starting in the Command Module, nicknamed Odyssey, traveling down through the tunnel into the Lunar Module Aquarius, and gave a close-up look at all the important systems that allowed a human being to walk on the Moon. It was a video tour de force, an unparalleled show for history and the ages. The TV session ended at 9:00 p.m. and Apollo 13 started doing more housekeeping, not knowing that almost no one outside NASA had watched their show. None of the three major networks carried their live broadcast. The public had lost interest. Quite a comedown from the days of Apollo 11.

Six minutes after the TV transmission ended, Astronaut Jack Lousma, serving as CapCom in Mission Control, instructed Jack Swigert to "stir up your cryo tanks," meaning check the quantity reading on the liquid oxygen tanks that were located in the unmanned Service Module. Checking the reading required Swigert first to stir the supercooled liquid electronically (it was in a so-called supercritical state, existing as something between a gas and a liquid) to ensure that it was uniform. He turned on the switch to do so.

Two minutes later Swigert alerted Mission Control, "Okay, Houston, we've had a problem here." Jack Lousma replied, "This is Houston. Say again, please," to which Commander Jim Lovell responded, "Houston, we've had a problem. We've had a main B bus undervolt." Thirty seconds after that Fred Haise reported, "We had a pretty large bang . . ."

When Swigert switched on the power to stir the tanks, two wires that had been accidentally damaged during a ground test several months earlier burned through their insulation and exposed the hot metal underneath. That caused a spark to jump between them, in a tank filled with pure liquid oxygen under extreme pressure. That in turn caused an explosion so powerful that it blew out the side of the Service Module, the force causing the co-joined spacecraft to spin erratically until the Command Module's small Reaction Control System thruster engines kicked in to stabilize it.

It was April 13, 1970, at 9:08 p.m. Houston time. The Apollo 13 Command Module was soon to be without electrical power, oxygen to breathe, water to drink, and the rocket engine required to bring them home. They were 178,000 miles away from Earth, more than three quarters of the way to the Moon, and they'd have to travel the rest of the way out there before they could start back.

It was almost midnight before the TV networks realized what was happening. CBS interrupted Merv Griffin's late-night talk show to let Walter Cronkite, roused from bed and rushed to the studio, break in to cover a live press conference from Houston.

I was asleep and didn't hear about it until the next morning, when Bob Manders called me at home. It was Tuesday, April 14. We switched on the TV and watched the morning news, overwhelmed trying to take it all in. By that time Lovell, Haise and Swigert had powered down the Command Module. Its backup battery had kicked in when the SM went dark, but that power was limited to a few hours at best, and much of it had already been used. What power remained needed to be reserved for reentry, if the crew of Apollo 13 made it that far. The astronauts had powered up the Lunar Module, crawled inside, and sealed the hatch between them and the co-joined Command and Service Module, the CSM.

They'd made a midcourse correction using the LM Descent Stage Engine, the first of many complex tasks that incredible Moon machine would be called upon to perform, all well beyond anything it had been designed to do. That thirty-one-second engine burn at 2:45 a.m. Houston time had put the spacecraft back onto a free-return trajectory.

The operation of the Lunar Module's hand controllers was reversed from those in the Command Module, and its guidance and navigation systems had never been programmed to maneuver the combined LM and twenty-nine-ton CSM into position for that burn. But the extensive training of the astronauts, and Fred Haise's encyclopedic knowledge of the LM systems, plus exhaustive analysis and simulations by the people in Mission Control planning the maneuver, had pulled it off. By the time I caught up, the astronauts had powered down the LM again to get what sleep they could manage. The Lunar Module power supply itself was designed to last forty-five hours with two passengers, plenty enough to take two of them down to the Moon's surface and back. Now it would have to stretch to twice that to get Apollo 13 back to Earth with three people aboard.

Although the spacecraft was on a course to whip around the Moon rather than get caught in its orbit, another engine burn would be required after Apollo 13 circled the Moon so that they could get home faster. It was a trans-earth injection burn that would put them on a path for reentry, except this time it would be done using the Lunar Module engine. It was scheduled for 8:40 p.m. Houston time.

We hadn't planned or simulated anything remotely like that, so it was time for me to go into work. I was no sooner at my desk when Dave Heath called; he'd been up pretty much all night. He wanted me to meet him in the Mission Control support room as soon as possible. I gathered up my notes and copies of our monitoring procedures and drove across the street. When I got there, the small room was already jammed. People were crowded around the table, stacks of accordion-folded computer printouts were stuffed under chairs, resting on top of filing cabinets, and spilling on the floor like Slinkys. There was a palpable tension in the room. A couple of engineers were working heads down, shoulder-to-shoulder, and another was doing pencil and paper calculations on a notepad. All of them were more than a little unsettled. I don't recall the names of anyone else in the room besides Dave; they were likely other contractor engineers and NASA employees. Dave spotted me and waved me in. I plopped my stuff on the table and sat down. He spent perhaps ten minutes filling me in and then pointed to another table with paper cups and a steaming coffee urn. "Pour yourself a cup," he said, "we're going to be here a while." He was right. From that Tuesday morning until splashdown on Friday, our team worked pretty much around-the-clock. I know I went home to get some sleep from time to time, but my sharpest memory is of the support room at the Mission Control Center.

Apollo 13's reentry into Earth's atmosphere would involve a series of spacecraft maneuvers that had never been attempted on any Apollo mission that preceded it. To help with the planning, the NASA team would need to know the angle within the narrow entry corridor at which the trans-earth trajectory would place them, and then determine the reentry path to splashdown that would result from it. To do that we had to know exactly where the Apollo 13 spacecraft was after the upcoming engine burn adjusted its orbit.

The power remaining in the Lunar Module batteries was being rationed, and there was considerable debate in the Mission Control room about whether to power up the LM fully again and use its guidance and navigation system to perform the trans-earth burn. The astronauts had been trained to use star sightings through the windows to triangulate their position. But Mission Control decided it was too risky to commit to this critical burn without the precise fix on location that the computer systems provided. The decision was made that the LM would be powered up for ninety minutes to allow the crew to perform the complicated series of steps required for the maneuver, and then shut down again. That meant the crew would be forced into even more draconian power conservation afterward.

On Tuesday, April 14, at 7:13 p.m. Houston time the Apollo 13 crew powered up the LM, activated its guidance and navigation system, and carefully positioned their spacecraft. Using newly developed procedures voiced up to them from the ground, they fired the LM descent engine for four minutes and twenty-three seconds while Jim Lovell used reverse hand controls to keep the ship aligned. When the firing was complete, the CapCom informed the crew that the burn was perfect—a welcome, if brief, taste of success for everyone involved. It was just before midnight in Houston, and for the moment everything was quiet. Lovell, Haise, and Swigert had powered down the LM again. Sleep was out of the question, but they were on their way back and needed time to rest. We got an update earlier on their new trajectory based on the tracking data and were given priority on the computers to run a reentry simulation. Their burn had indeed been perfect, and they would reenter right down the center of the entry corridor. We were hoping it stayed that way.

With sixty-three hours yet to go, a substantial toll had been taken on their spacecraft's power supply. If further engine firings were required, they were going to have to be done the old-fashioned way, manually. Midcourse corrections returning from the Moon were usually short ones, just long enough to adjust the spacecraft's path so that it stayed within the strictures of the entry corridor. On earlier missions, calculations of necessary adjustments to the spacecraft's trajectory had been done using the tracking data from a network of ground-based stations, and a series of star sightings by the astronauts themselves, using onboard instruments and careful measurements of viewing angles aided by the onboard guidance and navigation computer.

With the CM powered off, the LM powered down, and telemetry downlink from the spacecraft reduced to a trickle, Houston would need a lot of help from the crew to get Apollo 13 ready for another burn if required. To compensate for the lack of onboard systems and without access to the Command Module, Lovell would have to use sightings of the Earth's horizon and of the Sun through the LM window-mounted optics, with visibility greatly impaired by a cloud of explosive debris surrounding the outside of the spacecraft.

The three astronauts had to work together attentively to perform the burn correctly. Accuracy in exactly how the spacecraft was positioned for a midcourse correction was important, but with the Lunar Module heaters turned off to preserve power, the temperature inside the LM had dropped to 38 degrees, and all three astronauts were already suffering from the cold. To add to the distraction, Fred Haise had come down with an illness later diagnosed as a kidney infection.

The crew's much needed rest period had been short-lived. Not long after their nearly perfect descent engine burn, another serious problem cropped up. The carbon dioxide levels in the LM cabin air were building to a dangerous level, threatening the crew's safety. The CO_2 filtering cartridges had been designed for two astronauts' use on a lunar landing mission, not for the three of them on the long trip home, and they were getting overwhelmed. There were additional filters in the Command Module but the connections didn't match; it was a case of needing to fit a square peg into a round hole.

A team of Mission Control engineers worked up a "jury-rigged" solution using only materials that could be found inside the spacecraft. They voiced up instructions step by step to the Apollo 13 astronauts, who, in spite of their rapidly deteriorating physical health, painstakingly assembled it, pretty much the way it was portrayed in Ron Howard's 1995 film *Apollo 13*. It worked.

That sort of exhaustive and methodical work was done by so many people under these demanding circumstances. I'm writing this account as just one of those who were there. There were nearly two thousand NASA employees at the Manned Spacecraft Center at that time, most of them engineers, and many more contractor employees like me in offices located near the Center. All of them were laboring around the clock just like we were, using their unique skills and specialized disciplines to help bring Apollo 13 home.

The next test for the astronauts was preparing for reentry. Intersecting the atmosphere at an angle between 5.25 degrees and 7.5 degrees below horizontal was like threading a needle, but in this case the thread had unraveled, the sewing machine was damaged, and the tailors were unwell. No one wanted the Apollo 13 crew to have to do another mid-course correction burn.

Because of the ground tracking network, knowing exactly where Apollo 13 was at any moment, and how fast it was moving, wasn't one of our problems. During a normal, nominal, flight, the onboard computer would calculate the spacecraft speed and position and send it on to Mission Control. But there was also a more accurate and faster method in play. Throughout the Apollo Program, Mission Control communicated with the astronauts, interacted with the spacecraft systems, and received independent confirmation of its current location and velocity, the so-called state vector, through a worldwide network of specially equipped ground stations, oceangoing ships, and military aircraft, called the Manned Space Flight Network (MSFN). On the way out to the Moon and back, the state vector was also determined with the aid of NASA's Deep Space Network (DSN), three large dish radio antenna sites located at Goldstone, California; Madrid, Spain; and Canberra, Australia.

But speed and location weren't the only details those of us planning the reentry needed. We also needed an accurate prediction of where, and at what angle, Apollo 13 would enter the atmosphere; that is, the trajectory path it was following when it arrived. Once Apollo 13 got close enough to Earth, the Command Module computer would be powered up again, and since it would be flying the reentry, it would need to be brought up to date. Through the magic of rocket science, or the genius of the seventeenth-century German astronomer Johannes Kepler, once the tracking network got multiple fixes on Apollo's position and velocity, specialists at NASA and their contractors, including TRW, could accurately compute the trajectory of the new orbit it was following. That was how Mission Control had known that the earlier trans-earth injection burn was perfect.

On Wednesday morning, April 15, we got a trajectory update based on the latest tracking network data and saw that something had changed. Apollo 13's predicted path at entry had shifted upward; that is, away from Earth. They were going to come in at too shallow an angle and miss the corridor. We used the information to make another entry simulation run and confirmed the problem. We could not be certain that either the Command Module's onboard entry guidance or our recommended backup procedures would keep Apollo 13 in the atmosphere, let alone carry them to splashdown. We rechecked our data and ran the simulation a second time. Same result. All agreed that Apollo 13 would have to do another midcourse correction.

We couldn't figure out what had happened. The path had looked perfect after the last midcourse correction. Was the trajectory prediction wrong? ("Poppy" Northcutt, our TRW analyst who had computed it, knew her stuff. It wasn't). Or was something screwy on Apollo 13 itself? The Service Module had exploded—who knew what was going on back there? At that point, none of the speculation as to "why" mattered;

Apollo 13 would miss the entry corridor if something wasn't done. The next midcourse correction burn, the one we all hoped not to do, was scheduled for later that night. There was nothing more we could do but wait.

At 10 p.m. Houston time the crew briefly powered up the Lunar Module to prepare for the burn. Because of the tightfisted apportioning, the onboard batteries now had some power in reserve, barely enough to allow for the burn to be done manually and still get Apollo 13 home. They were still about thirty-six hours from reentry, so turning on the rest of the navigation systems was out of the question. Jim Lovell rotated the LM until he could align the top of his optical sighting with the edge of the Earth. That maneuver positioned the spacecraft "heads-down" relative to the way the spacecraft was traveling, so that firing the LM descent engine would push their trajectory deeper into the Earth's atmosphere. Lovell and Haise used their hand controllers to keep the LM in place while Swigert counted down to the firing. At 10:31 Lovell fired the LM descent engine at 10 percent thrust and shut it down on Swigert's mark fourteen seconds later. At 10:33 the crew powered down the LM systems.

With their drained health, it's hard to imagine the mental and physical exhaustion that effort must have cost them. But once again they rose to the task. Later that evening we received Poppy's revised calculation of Apollo 13's predicted angle in the entry corridor. It was indeed steeper. We ran another reentry simulation and found that both the CM guidance computer and the crew backup procedures would carry them safely to splashdown. We didn't know what had caused the shallowing in the first place, but the midcourse burn had fixed it. The astronauts were back out of the woods, or maybe farther into the clearing.

Things brightened further during the early morning hours on Thursday, April 16. The Spartan management of the spacecraft's power reserves had gained Apollo 13 a sufficient margin so that the Lunar Module would now reach Earth with a little to spare. If somehow the Command Module's batteries could be recharged a bit, and the loads on them then carefully managed, the CM batteries could be used as a source of power during the last six hours before reentry. Apollo's Command Module and Service Module systems had been designed to

The complex reentry procedure required for Apollo 13's return to Earth, April 17, 1970. NASA Illustration.

transfer power between the two spacecraft just one way, from the Command and Service Modules to the Lunar Module, and then barely at a trickle. Undertaking the challenge, the engineers on the ground, with the help of John Aaron's leadership as EECOM, worked out a way to reverse the process, to use the now surplus battery power in the Lunar Module to recharge the Command Module batteries. They walked the crew through the steps, and once again they succeeded. The LM was still running on its own limited power while it also recharged the CM batteries, so the process would take the rest of that day, sixteen hours in fact. But prospects were definitely looking brighter for the crew.

The ground tracking network continued to provide updates during the day. The predicted trajectory path into the atmosphere had shallowed a little again—puzzling, but not yet alarming since the onboard computer could handle it. Our teams ran another set of trajectory and reentry simulations and adjusted the data to be voiced up to the astronauts the next morning so that they could key it into the Command Module guidance programs and the entry backup system and initialize them correctly for this reentry: the specific combination of entry location over the Earth, angle within the corridor, and speed at the start of reentry.

Meanwhile Mission Control and the crew had their hands full. Before Apollo 13 could return to Earth, the astronauts would have to perform an unprecedented and complex set of maneuvers to position the Command Module for reentry. First, they would power up the Command Module and check out each of its systems. Then the Lunar Module and the Service Module, neither of which could survive a fiery return, had to be jettisoned away from the Command Module, each in a different direction to get them out of the way of the CM and of each other. Finally, Jack Swigert would rotate the CM heatshield forward for reentry. This was a wholly new and taxing procedure, complex and never rehearsed, that required more than four hundred separate crew actions, each of which had to be tried first in the simulator by astronauts on the ground and then voiced up and confirmed one step at a time by the exhausted crew. Apollo 13's original but now grounded Command Module Pilot, Ken Mattingly (bumped from the flight over concerns that he'd been exposed to German measles), had a role in developing this complicated sequence, one of the ways his knowledge of the Command Module and his experience training for this flight with Lovell and Haise paid dividends.

Sometime around noon on Thursday, with the astronauts still a day away, the mission measurements showed that Apollo 13 had gone off course. A new ground tracking network projection showed Apollo 13 had drifted up in the entry corridor yet again, this time to minus 6.1 degrees below horizontal—and later still it changed to minus 6.0 degrees. Left uncorrected, the Command Module would brush only the top edge of the corridor, barely within the capability of the guidance computer to manage it, and even then with a far-too-elongated ride through the thin upper atmosphere, flying lift-vector (i.e., the tilt) down, until the g-forces finally built up enough to slow down the Command Module. Nobody had answers as to what was causing this ongoing upward drift, so Mission Control was not going to risk it. They informed the crew and scheduled the burn for 7:00 a.m. Houston time the next day, five hours before reentry.

There was a little good news. Though the Lunar Module engines would again be called on to perform the burn, the Command Module batteries had recharged sufficiently that the guidance computer could be powered on to help guide the crew through the process. Lovell would have to repeat his routine of using the Sun and the edge of the Earth to position the Lunar Module optically, and he and Haise would fire the engines and hold the spacecraft steady while Swigert called down readouts of the spacecraft's attitude from the CM computer. The other piece of good news was that the burn would be a short one, so it could be done using the LM's four clusters of small Reaction Control thrusters rather than the big Descent Stage engine. Immediately after the burn, most of the CM systems would be shut down again to preserve all power possible for the complex maneuvers coming up just before entry.

In Houston we spent the rest of that day listening to Mission Control communicating with the astronauts while we reworked the data the Apollo entry guidance programs and backup system might need once the Command Module was back in the entry corridor. If any updates were required, they'd be voiced to the crew before the CM was fully powered up for reentry.

Friday, April 17—reentry day—started early for the crew of Apollo 13. At 2:48 a.m., about nine hours before reentry, they powered up the Lunar Module and used its power to power up the Command Module briefly, to test its systems. To everyone's relief the long cold sleep had not damaged them, and the CM batteries were fully charged. Next, they transferred all necessary equipment and supplies from the LM to the CM and began setting up for the midcourse correction burn.

At 6:53 a.m. Lovell, Haise, and Swigert pivoted the spacecraft and fired the LM RCS engines for 21.5 seconds. The entry angle was nudged toward the center of the corridor. Not long after that we got a new trajectory update, and they were back where they needed to be. That had to be enough; there wasn't time to do another burn. Twenty minutes later Apollo 13 began the most complex and risky dance of spaceships in the history of human space travel. The last four hours before reentry required the fatigued, dehydrated, and ailing crew to perform a complicated and unrehearsable series of spacecraft maneuvers.

Neither the Lunar Module nor the Service Module was designed to withstand the heat of reentry. For Apollo 11 and Apollo 12, the Lunar Module was discarded at the Moon, and on Apollo 9, the only other time it was flown near Earth, it was abandoned in orbit. On all previous Apollo missions the astronauts had used the Command Module to orient the Service Module away from their own line of travel, fired the explosive bolts to separate the SM from the CM, and used the SM's own engines to move it a safe distance away. The SM burned up during reentry, and any fragments dispersed safely into the ocean.

None of that would work this time. The LM would have to do the bulk of the work before powering up the CM, and because of the aborted flight plan, an additional problem had arisen before the LM could then go on its way. Apollo 13's Lunar Module included an experiment intended for use on the Moon that was powered by a small nuclear generator. The experiment would have been left behind there, but as a contingency, its plutonium oxide core was enclosed in a specially made ceramic cask that was designed to withstand a violent explosion or a fiery reentry. But now the Atomic Energy Commission contacted NASA to insist that the final reentry trajectory of the Lunar Module must be directed to ensure that any of the experiment's parts that did survive entry would plunge into the deepest part of the Pacific Ocean. Seriously! Part of the need for putting the astronauts through the arduous four-hundred-plus-step procedure to prepare for reentry was dictated by the AEC.

As soon as the last midcourse correction was finished, the crew got busy. Lovell and Haise crawled back into the LM and, with the help of its abort guidance system, oriented and held the spacecraft into position to jettison the Service Module. Swigert powered on the Command Module systems to let him fire the explosive bolts connecting the SM to the CM. Lovell ignited the LM engines and gave the stacked spacecraft a forward push; Swigert detonated the bolts and Lovell immediately fired the LM engines, backward this time; and the two now-separated spacecraft moved apart in opposite directions. The Service Module's momentum carried it away from the Command Module and onto a separate trajectory to reentry.

As the Service Module drifted past the LM window, Lovell and Haise were able to watch it and take a few pictures. Lovell was stunned by what he saw. "There's one whole side of that spacecraft missing," he reported. Haise added, "It looks like it got the engine bell too—man, that's unbelievable." The Service Module had suffered an enormous explosion, later estimated to have been as powerful as seven pounds of TNT, and it was utterly destroyed. The blast had originated near the forward end of the Service Module, not far from the Command Module's heatshield.

Swigert turned on all the remaining systems in the Command Module. The CM was using its own batteries now, so power transfer from the LM to the CM was stopped. Swigert and Haise checked out all systems one by one while Lovell remained in the LM to prepare it for separation. Swigert signaled that the Command Module Guidance Computer program was activated and ready for reentry—no last minute updates from our team had been necessary—so Lovell flew the Lunar Module one last time. He used the LM Reaction Control thrusters to rotate the co-joined spacecraft into the prescribed separation position and then allowed the reactivated LM Abort Guidance Computer to hold it there. He worked his way back up through the tunnel, sealing the hatches at each end and lowering the pressure in the space between them to about 3.5 pounds per square inch. He joined Haise and Swigert in the CM.

On previous flights the Lunar Module was jettisoned by unlatching it from the CSM and firing the SM RCS thrusters to separate the two. The Service Module was gone now, so Mission Control had worked up a different way to do that too. When the Command Module undocked from the Lunar Module, exposing the 3.5 psi pressure in the connecting tunnel to the vacuum of space, the difference in pressures would act like a small rocket firing. At 10:43 a.m. Swigert uncoupled the two ships, and that's exactly what happened. Apollo 13's Lunar Module moved away from the Command Module and on toward an intersection with the atmosphere. It ended its memorable journey, and its unplanned but successful duty as a lifeboat, in a fiery pyre that made even the Atomic Energy Commission happy. (Apollo 13's ceramic cask of inert plutonium oxide is now safely entrenched four miles deep in the

Pacific Ocean's Tonga Trench. The cask is expected to withstand sea water corrosion for more than eight hundred years, much longer than the half-life of the plutonium itself.)

"Houston, LM jettison complete," Swigert reported.

"Copy that," said CapCom Joe Kerwin, "Farewell, Aquarius, and we thank you."

Swigert used the CM Reaction Control System thrusters to rotate the Command Module fat-end forward, and the Entry Guidance Computer program took over. The crew were on their own. The only part that my team and I could contribute now had been written and delivered to them two months earlier.

That last midcourse burn put Apollo 13's predicted angle of entry into Earth's atmosphere at 6.5 degrees; that is, sloping 6.5 degrees below the horizontal toward the Earth. At 36,200 feet per second, almost 25,000 miles per hour, the speed Apollo 13 would be moving at that point, would put them right on target for a nominal reentry flight of 1,285 nautical miles to splashdown in the Pacific. That was one of the so-called standard Apollo reentry trajectories for which I'd run computer simulations many times before. I'd even flown that entry myself in NASA's Command Module Simulator. I knew how the onboard computer would navigate that roller coaster, the displays that Jack Swigert would keep an eye on to make sure it did things right, and which of our backup procedures he would follow if he decided to take over control.

Now I could only wait while Apollo 13 began its unavoidable period of radio blackout. The exact moment Apollo 13 entered Earth's atmosphere, the so-called Entry Interface (EI), was a mathematical calculation, not a physical event. EI was defined, somewhat arbitrarily, to be at an altitude of 400,000 feet. At that height there's no actual atmosphere, and it is generally considered "outer space" by the scientific community. In the early days of human spaceflight an astronaut was said to have "earned his wings" if he reached an altitude of 50 nautical miles (264,000 feet). But setting Entry Interface at 400,000 feet gave the crew time to do a final checkout of their onboard instruments and to focus on the guidance entry program as it executed its first few steps.

The actual "flying" began twenty-eight seconds after EI, when the Command Module encountered a backward force of 0.05-g pressing

NASA Mission Operations Control Room team awaiting the end of Apollo 13 communications blackout, April 17, 1970. NASA Image.

against the spacecraft as it flew. Not much of a resistance, but strong enough that the ship's sensors could detect it, and at that speed, friction enough to start boiling away the heatshield. That event signaled to the crew that they were at the edge of the "sensible atmosphere," where there was enough air density to push back on Apollo and begin to slow it down. From that point on, the roller coaster ride was under way.

As the Apollo 13 Command Module pushed deeper into the atmosphere, the g-force pushing back on it continued to rise, peaking at more than 5-g, meaning each astronaut would feel a 750-pound weight pressing against him. The friction blasting the Command Module's heatshield caused temperatures outside to spike to more than 5000°F. As the Command Module burrowed its way through that inferno, the heatshield would slowly and continuously melt and blow away, transforming its outer surface coating directly into a gas, a process called ablation. That would carry away with it much of the heat and protect the spacecraft and its occupants. That extremely hot and electrically charged gas surrounding the Command Module during reentry, basically the world's worst radio static, prevented any communications with the crew for about three and a half minutes, until the outside surface of the spacecraft cooled down again. By that time the heatshield, if it held, would have exhausted its usefulness. If the explosion in the Apollo 13 Service Module had also damaged the heatshield, the crew's fate would be determined while they were out of contact with Mission Control.

The three and a half minutes came and went with no communication from the crew. That was a little troubling but not alarming. This was, after all, anything but a nominal flight. At four minutes of silence, I was becoming concerned. When another thirty seconds went by with no word from the crew, my hopes sank. I remember the awful knot in my stomach; my chest tightening with restrained grief. I looked at Dave and then at the others around me and I could see we were sharing the same thought. Nobody spoke but we all knew—we'd lost them. After all everyone had done to help these men survive, after the extended NASA team around the world had come together to work a miracle, it wasn't enough. The one piece of hardware we couldn't work around, a damaged heatshield, was all that mattered. I was young then. I hadn't lived for enough years to experience such a crushing reversal at the hands of a disinterested universe, let alone comprehend it.

And yet, if you know your history, or if you've watched Ron Howard's movie, you have an advantage that we didn't. You know how it came out.

At five minutes after the start of reentry blackout, and after a seeming eternity of hopelessness, we heard Jack Swigert's voice come over the speaker and say, "Okay, Joe." To which Joe Kerwin responded, "OK, we read you Jack." And that was that. The crew of Apollo 13 were safe.

Life had other taxing moments in store for me, but nothing to come matched the emotional swing I felt that day. No future heartbreak has ever carried its own redemption so closely in trail. There's a phrase that has been worn to meaninglessness by overexposure to underwhelming events, but it was precise that day. It left me breathless.

On Friday, April 17, at seven minutes past noon in Houston, the Apollo 13 Command Module, with Jim Lovell, Fred Haise, and Jack Swigert on board, splashed down in the Pacific Ocean just four miles away from their recovery ship, the USS *Iwo Jima*. Forty-five minutes later they were standing on its deck.

Why had reentry blackout lasted so long? The answer came much later, after we got a chance to examine more of the data that showed the Command Module's actual reentry angle when it encountered the atmosphere. After that final midcourse correction just five hours before reentry, the spacecraft had moved up in the entry corridor yet again. When Apollo 13 reached Earth it was still in a slightly shallower position than anyone had predicted. Fortunately, the spacecraft was immersed deeply enough to slow the Command Module and let the density of the atmosphere take hold. Then the onboard guidance computer began the long and bumpy ride that carried the astronauts home. Since the CM had encountered thinner air at the start of reentry, the computer likely had to push deeper and longer than planned to slow down the spacecraft and, once it gained enough traction, to allow the heatshield to burn away and cool itself down. That part of the flight took a lot longer than anyone had accounted for: one and a half minutes longer, in fact. When it finished that task, the guidance program flew the Command Module on a wholly new roller coaster ride and successfully navigated Apollo 13 to splashdown right next to the U.S. Navy carrier that was waiting for it.

To get a sense of that, a nominal entry flight for Apollo 13 would have lasted fourteen minutes and flown 1,285 miles until splashdown. The actual reentry did take fourteen minutes, and the Command Module opened its chutes close enough to USS *Iwo Jima* that the crew in the recovery helicopter watched it descend. But it had to spend an additional one and a half of those fourteen minutes in the scorching heat of blackout, fighting to get a bite on a too thin atmosphere, before it won the battle and then came home with a bullseye. Those one and a half minutes are still the longest of my life.

The crew of Apollo 13 had returned safely to Earth. In spite of the horrendous force and timing of the Service Module explosion, there were, in the end, several factors that went Lovell, Haise, and Swigert's way. Most important was the fortitude of the astronauts themselves, performing herculean tasks under punishing circumstances. Then came the unparalleled performance of the Grumman-built Lunar Module: when pressed into service for responsibilities well beyond anything conceived during its design, it rose to the challenge. Add to that the NASA team's unrelenting determination to bring the crew back alive. The result was a resounding success and the crew was rescued. I believe it's as exceptional an achievement of human determination and resourceful performance under fire as was Neil Armstrong's first step onto the Moon.

Well after the celebrations of Apollo 13's safe return settled down, I got feedback from Dave Heath of an idea someone had proposed as to why Apollo 13's entry trajectory kept shallowing. The gaping damage in the hull of the Service Module was leaking propellants, the theory went, and it was outgassing who-knows-what, all the way to the Moon and back. This material spewing from the side of the SM must have acted like a small rocket pointed in that direction. It gave the co-joined spacecraft a small but constant sideward thrust that in turn pushed the trajectory ever so slowly upward, away from the Earth.

I repeated that story many times, until I had a chance to talk to John Aaron about it a few years ago. It turns out that theory was wrong. The problem came not from Apollo 13's crippled Service Module but from its intrepid lifeboat, the Lunar Module. John said that SM outgassing probably had little effect. The largest contributor to the shallowing was caused by a continual stream of water vapor venting sideways outside the Lunar Module from the overworked water system the three astronauts had depended on for drinking and cooling during their 240,000-mile journey home. I never stop learning something new about Apollo.

There was a personal footnote to the rescue of Apollo 13. Jim Lovell, Jack Swigert, and Fred Haise returned safely to Earth on April 17, 1970, which was a Friday. I was pretty much fried, to use an expression we didn't have back then, and spent Saturday at home sleeping, getting further acquainted with my young son, and little else. On Sunday morning Barbara and I went to a late mass ministered by our pastor, the same priest who'd informed us during midnight mass on Christmas Eve of 1968 that Apollo 8 was on its way home. This time during his sermon he shared a unique perspective of Apollo 13.

"Many people have told me how hard they prayed for the Apollo 13 astronauts last week," he started. "I agreed with them and said, so have I. They said how thankful they were that God had answered their prayers and brought the crew home safely."

The pastor paused for a moment and looked at the congregation.

"I watched television coverage of Apollo 13 whenever I could," he said, "and I listened on the radio the rest of the time. At no point did Mission Control report that a giant hand had reached down from heaven to lift the Apollo 13 spacecraft in its palm and carry it gently back to Earth." He paused again and then said, "God doesn't work that way. He answers our prayers by giving each of us a brain to make use of, a capacity to learn, the intelligence to apply it, and the free will to choose to. After that it's up to us." He slowly swept his outstretched hand across the congregation. "You chose to use your God-given gifts this week, you and the others around the Earth who gave it everything they had to bring Lovell, Haise, and Swigert home. Each of you, and all of you, saved Apollo 13. You are the ones who answered those prayers."

Apollo 14: Nearing the End of the Road

Of all the people I worked with at TRW, Charlie Porter had taken the most unlikely route to become a member of the Apollo Program team. He had a degree in aerospace engineering from the University of Texas, but he was also a professional water skier; he was a part of a ski team that performed around the world—from the United States to Japan, the Philippines, and Burma—doing doubles skiing, pyramid skiing, slalom skiing, barefoot skiing, and acrobatic ski jumps.

In the summer of 1970 he decided to narrow his focus to the more sedate life of putting men on the Moon and came to work at TRW. He joined the reentry planning group just in time for me to get him ready for Apollo 14. His adroitness at waterskiing must have helped; he jumped in with both feet and got to work. There was a known bug in the guidance software running the Real Time Computer Complex reentry simulation program, and we'd been making adjustments to our computations to account for it. Charlie wasn't satisfied with that; he applied his engineering talent and our Friden mechanical calculator to design a way to work around it. From the start, I knew Charlie was going to work out just fine.

It was midsummer and the launch of Apollo 14 was still six months away. Unlike the unrelenting pressure of preparing for the previous flights, every waking moment now wasn't about work. When I learned

Apollo 14 Crew (*left to right*): Stuart A. Roosa, Alan B. Shepard Jr., and Edgar D. Mitchell, 1971. NASA Image.

that Charlie was a skier, I asked him to teach me and he was happy to agree. The torpid and pollution-laden "Clear" Lake was a far cry from his ski shows at Naha Harbor in Japan, and while our locals were cautioned not to swim in those waters, the lake had a thriving sailboat rental business and a ski jump ramp for the less faint of heart. Another TRW engineer, Joe Alari (whose Cuban parents had fled to the United States when Fidel Castro came to power), owned a small powerboat, complete with a tow rope, and the three of us set out on a summer Saturday to challenge the lake. I took a lot of spills that day, closing my mouth and holding my breath to keep from swallowing the water, and took even more spills the following weekend, but eventually I learned how to stand up and stay standing. Joe and I took turns piloting and getting towed around the lake while Charlie shouted encouragement to us from the stern.

Charlie was a great coach; by the end of summer I was skiing slalom, keeping both feet on a single ski. I was feeling pretty good about myself, so I asked Charlie if we could try the ski ramp. He had a pair of jump skis, stronger than regular skis with undersized rudders, to navigate the smooth, flat wooden surface of the ramp. Charlie insisted that we wear flotation jackets, and he showed us how to keep our head up and bend our knees as we approached the ramp. He had Joe tow him over the jump first to show us how it was done; he made it look easy.

I don't remember whether Joe or I went next. I fell on my first try and lost my skis, and I forgot that Charlie had said let go of the tow rope. I was dragged face forward over the ramp. The wood was slimy with algae so nothing was injured but my ego. I righted myself and shouted to him that I wanted to try again. My second jump was much better. I hit the ramp standing, stayed standing all the way to the top and over, and surprised myself by landing upright as my skis slapped the water. I did it, and it was a rush. This was my chance to quit with my success untarnished. Instead I shouted, "Let's do it again!"

Joe towed me in a wide arc and pulled me back toward the ramp. In my overconfident zeal I promptly forgot everything Charlie had taught me. Halfway up the ramp I toppled over, but my skis didn't come off this time, and once again I didn't release the rope. I slid sideways up the ramp and nosedived five feet into the lake. My legs buckled when I hit the water, the skis crossed and slammed edgewise into my knees. The pain was so intense I couldn't move. If it weren't for the buoyancy vest I would have drowned. Joe pulled the boat around and he and Charlie lifted me out of the water, my legs streaming blood. I was lucky; I had deep purple bruises and stiff knees for the next couple of weeks, but nothing was broken. Charlie and Joe went skiing every chance they got after that, but I put my waterskiing days behind me for good.

Apollo 14 lifted off on January 31, 1971, bound for the Moon's Fra Mauro Highlands, with Commander Alan Shepard, Command Module Pilot Stu Roosa, and Lunar Module Pilot Ed Mitchell onboard. Shepard was a Naval Academy graduate and the first American to travel into space in 1961. Roosa and Mitchell were both making their first spaceflight. Stu Roosa was an Air Force officer who earlier in his career had been a U.S. Forest Service "smokejumper." He'd parachuted into remote and inaccessible areas of California and Oregon to fight wildfires. Ed Mitchell was another Naval Officer, also an aeronautical engineer and test pilot, but that's where he broke with the mold. Even as a NASA astronaut he professed belief in extrasensory perception and paranormal phenomena, and he asserted that UFOs are in fact alien spacecraft. Pretty far afield from what I learned in engineering school, but maybe that's another reason why I favor the Command Module Pilots.

At age forty-seven, Alan Shepard was the oldest astronaut to walk on the Moon. After he became the first American in space in 1961, when he rode his Mercury spacecraft on a fifteen-minute suborbital flight, he was grounded until 1969 by an inner-ear condition that was later corrected by surgery. He was subsequently assigned to be Commander of Apollo 14, and on February 5, 1971, he and Astronaut Edgar Mitchell became the fifth and sixth humans to walk on the Moon.

Besides his age, there was another notable achievement on Alan Shepard's mission to the Moon. Shepard became the first and so far only golfer on the Moon. In his personal gear he'd stowed two golf balls and a 6-iron club head that had been custom designed by a golf pro to fit onto the handle of a soil scoop. Shepard wanted his golfing session to be a surprise, but he was not so foolish as to ignore asking permission from the Director of the Manned Spacecraft Center, Robert Gilruth. Gilruth adamantly refused, but Shepard finally wore him down by promising his "stunt" would not interfere with any of the science or other activities that were planned for his flight's already tight schedule. By the end of his second Moon walk, Shepard needed all the schedule flexibility he could squeeze in to allow him the extra time to make his golf shot, but he was determined and got there.

Because his spacesuit was bulky and not very flexible, he had to swing the makeshift golf club one-handed. He dropped the first ball, whiffed at it twice, and then hit it short. But he made solid contact with the second one and off it went. He claimed he drove that ball "miles and miles and miles" in the low gravity and airlessness of the Moon, but later photographic analysis showed it went about three hundred to four hundred yards. Still, an impressive shot with a 6-iron.

I was still the Task Manager for TRW's Apollo reentry work for Apollo 14, but I was out of the day-to-day support during flight. Charlie Porter had quickly risen to that task and was also doing the reentry planning for Apollo 15. My assignment now read "responsible for

Apollo reentry efforts performed by the team" and "perform special studies." Some of that meant I'd be doing a post-flight analysis of how well the Apollo 14's Onboard Entry Guidance Program had performed during the mission. I did the work, wrote up a report, and sent it to NASA a couple of months after splashdown. That didn't consume all of my time, so I spent the rest of it on those "special studies," which really meant helping TRW get new business from NASA.

After budget pressures forced NASA to cancel the last three Apollo launches, only Apollo missions 15, 16, and 17 remained. The last of those flights was scheduled to launch at the end of 1972, and then only if there were no further budget cuts. TRW, and therefore I, would be out of work for NASA in two years, so our company wanted to do something about that. There were two new space programs in NASA's plans in which we could try to play a part. One of them was called Skylab, an early test of an orbiting space station. The program would use existing Apollo Command and Service Modules from the canceled missions to send astronauts into orbit. A repurposed Saturn IVB rocket casing would serve as the orbital laboratory. NASA's other new program was just coming off the drawing boards: an entirely new manned spacecraft with reusable launch engines and a winged rocket-plane perched on top of them. They had named it the Space Transportation System. Except for two weeks in July 1971, when I did some analysis of Apollo 15's entry guidance programs, and later when I attended a crew briefing for Apollo 17, I was pretty much finished with Apollo. I'd spend most of the next three and a half years at TRW trying to get us some new business.

Apollo 15: Phil Shaffer

Though I didn't play much of a role on Apollo 15, the little I did afforded me a learning experience I wouldn't soon forget. It came courtesy of another memorable NASA professional named Phil Shaffer. Phil had been a Flight Dynamics Officer and a Flight Director on the Apollo missions up through Apollo 11, meaning he was one of the folks who sat at a console in the Mission Control Room during a flight. By the time I encountered him, he was Assistant Chief of the NASA Houston Flight Dynamics Branch. During one particularly painful meeting with

Apollo 15 Crew (*left to right*): David R. Scott, Alfred M. Worden, and James B. Irwin, 1971. NASA Image.

him, Phil schooled me on the use of good judgment. I was destined to have more meetings with Phil later on, during my work on the Space Shuttle. What I learned from him that day on Apollo 15 came to my rescue the next time.

In the run-up to each Apollo mission Phil conducted a weekly readiness review session in an auditorium-sized room at the Manned Spacecraft Center. The NASA staff responsible for the various elements of an Apollo mission, plus their support contractors, were expected to update Phil on the progress of their work preparing for the upcoming mission. These meetings were uniformly dreaded.

Phil was physically intimidating. He stood well over six feet tall and carried about three hundred pounds, very little of it baby fat. A colleague of mine once described sitting in a conference room when Phil arrived. Though his back was turned, he knew when Phil got there because the light from outside in the hall was blocked as Phil came through the door.

Phil had a low tolerance for bullshit. During his sessions he'd sit at a table at the front of the room, facing his audience, displaying neither a smile nor any other pleasantness. He was an unrepentant chain smoker, so a veil of smoke continuously framed his face, as if reinforcing some inner smolder waiting to erupt. It's hard to envision now just how much people smoked in meetings in those days.

When someone presented to him, Phil would listen without comment, except for an occasional question. If he was satisfied, he said "OK," consulted his agenda, and asked for the next presenter. If he wasn't satisfied, what Phil did next didn't bode well for the presenter. I was on the receiving end of one of those responses.

In an earlier session Phil had asked my NASA contract task monitor Dave Heath to verify the value of a particular onboard computer parameter, called "Minimum UPCONTROL Altitude." It was one of the specific computer calculations the Apollo Command Module Pilot monitored during reentry to be sure the guidance computer was functioning correctly. The details aren't important other than a value of forty nautical miles was the number currently in use. Dave had handed

Phil Shaffer, Assistant Chief of NASA Flight Dynamics Branch. NASA Image.

off the task to me and, always eager to please, I set about it with my typical obsessive, overly thorough approach. When the time came to report back to Phil, I was given the opportunity to do so. I took that as a vote of confidence, but later I wondered if Dave, who was a battered veteran of Phil's sessions, had little appetite for presenting it himself.

I came to the meeting armed with a half dozen very detailed charts that I'd prepared. When I was called on, I put the first one on the overhead projector and began talking Phil through the in-depth mathematical analysis I'd done. When I got to the second chart, Phil took the cigarette out of his mouth and rested the hand holding it on the table. He stared at me, frowning but not saying anything, while the cigarette burned itself down.

I broke into a sweat. It didn't look as though Phil was too happy, so I kept on talking, hoping to postpone the time he'd let me know. As I put up the third chart, Phil scanned it, glanced at the stack I had remaining, and raised his cigarette hand, palm out.

"Stop," he said.

I fumbled a bit but quit talking.

"Is there a point to this?" he asked me.

"I . . . I was trying to show you how I calculated the minimum altitude."

"Put down the charts. Just tell me—is forty good or not?"

"Uh . . . it's not that simple. It depends on the time of year, the latitude of the reentry point, the angle of—"

"Just tell me the goddamn minimum."

I swallowed my breath. It took me a second or two to recover. I shuffled through my charts, put the last one on the projector, and pointed to it.

"Under worst case conditions, the minimum altitude should be forty-one nautical miles, not forty."

Phil stared at me as if he'd discovered a special breed of idiot.

"Christ," he said, not quite inaudibly, and then after a pause, "We're not that good."

I wasn't sure what to do next. Phil looked over at Dave, who was seated nearby. "You're recommending no change." He made it a direction, not a question.

Dave nodded and said nothing.

"That's how we'll record it," Phil said.

He put his cigarette in his mouth, made a note, and looked up at me again. "Sit down." He glanced at his agenda. "Who's next?"

I sat.

Two years later, after the Apollo Program was officially over and I had moved on to work on the Space Shuttle, I was to have another encounter with Phil. That one would go a little better.

Apollo 15 lifted off on July 26, 1971, with Commander Dave Scott, Command Module Pilot Al Warden, and Lunar Module Pilot Jim Irwin aboard. Dave Scott was cut from the traditional astronaut cloth: Boy Scout as a kid, West Point graduate, Air Force test pilot. This was his third spaceflight; he'd flown with Neil Armstrong on Gemini 8 and with Jim McDivitt and Rusty Schweickart on Apollo 9. On this flight he carried a plaque engraved with the names of eight astronauts and six cosmonauts who had died in the pursuit of space travel. He left it on the Moon as a memorial to them.

Jim Irwin and Al Warden were making their first flights. Jim Irwin was a Naval Academy graduate and had a Master of Science degree in aeronautical engineering from the University of Michigan. Al Warden, like Dave Scott, graduated from West Point, but like Jim Irwin, he held a master's in aeronautical and astronautical engineering from Michigan.

Scott and Irwin spent sixty-six hours and fifty-four minutes on the Moon while Al Warden orbited above, a record at the time but later broken by both Apollo 16 and Apollo 17. However, the greater distance between the Earth and the Moon on that particular flight, and therefore between the Earth and Al Worden while he orbited the Moon, earned him an official entry in the Guinness Book of World Records as "most isolated human being."

The Apollo 15 Moon lander touched down in the Hadley-Apennine lunar region on July 30, 1971. This mission was the first to use the Lunar Roving Vehicle, a battery-powered "beach buggy" that allowed the crew to explore and perform scientific experiments over a range of more than seventeen miles and capture spectacular photos and video along the way.

Apollo 16: Distant Crackling Thunder

The event that most dramatically highlighted the end of my time on the Apollo Program was not the last flight to the Moon on Apollo 17 but the one that preceded it. In late 1971 I was transferred to the TRW offices in Cape Canaveral, Florida, to work on a small Space Shuttle–related task that our Florida organization had received from one of the NASA directors at the Kennedy Space Center. The assignment was short-lived—I was there less than a year—but my time at KSC provided me with one extraordinary benefit; I witnessed an Apollo launch up close and in person.

At 12:54 p.m. on April 16, 1972, Apollo 16, bound for the Moon atop a thundering Saturn V rocket, carried astronauts John Young, Ken Mattingly, and Charles Duke up into a stunningly blue Florida sky. NASA had made provision for their employees and contractors, plus some of our family and friends, to watch the launch from a narrow beach fronting a causeway inside the Kennedy Space Center property. I invited fourteen people, including some from as far away as Texas. It's amazing how many folks remember they're your closest friends when they find out you have an invitation like that. We spread our blankets on the sand and had a clear view across the salt marshes to Launch Complex 39A, about five miles distant.

A Saturn V launch is like everything you've read. A sudden silent flash of orange at the base of a tall white spire, followed seconds later by a muffled boom and the beginnings of a distant crackling thunder that could be felt as well as heard. The air surrounding the rising spaceship, concussed by the 7.5 million pounds of thrust coming from the Saturn V's huge rockets, buffeted our eardrums as it washed over us. Apollo 16 went straight up, moving slowly at first as if testing its footing, then it passed behind a scattering of cumulus clouds, seemed to shift gears, and arced up and over and away like a force of nature, trailing a blazing cylinder of fire and an elongated white contrail.

I've seen other launches firsthand since then, from the eerie, sunset launch of an unmanned Air Force mission that rose overhead as I stood in my backyard in Merritt Island, Florida, to two launches of the Space Shuttle that I witnessed at KSC during my Shuttle days, to a multistage

Apollo 16 Crew (*left to right*): Thomas K. Mattingly, John W. Young, and Charles M. Duke Jr., 1972. NASA Image.

night launch of an unmanned spacecraft lifting off at dusk from NASA's Wallops Island launch pad in Virginia. Each of these was spectacular in its own way, but nothing equals experiencing the immense power and emotional import of the Saturn V rocket propelling three human beings into space on their journey to the Moon. Only twenty-four people have reached the Moon, all of them Americans, and of those, twelve walked on the surface. I had the opportunity to be present when three of those voyagers set out on Apollo 16. I felt privileged to watch them go.

Commander John Young was on his fourth flight into space, a veteran of two Gemini flights and now on his second Apollo mission to the Moon. Command Module Pilot Ken Mattingly was making the first of his three space flights. (He'd originally been slated to fly on the ill-fated Apollo 13; his next two would be on the Space Shuttle.) Lunar Module Pilot Charlie Duke was making his first and only flight. He was thirty-six years old when the Apollo 16 Lunar Module touched down, making him the youngest person to walk on the Moon. Eight months later, on Apollo 17, Harrison Schmitt bested his record by three months.

On April 22, 1972, John Young brought his Lunar Module to a landing spot on the Descartes Highlands and became the ninth person to walk on the Moon. He was a naval officer and former test pilot, and in an iconic photograph taken during his spacewalk by his crewmate Charlie Duke, Young showed his enthusiasm for the mission, and for the country he served, by literally acting out Neil Armstrong's "giant leap for mankind." He leapt from the lunar surface and saluted the American flag while still in mid-"air." The photo taken by Charlie Duke of John Young hovering next to the American flag, with the Lunar Module Orion and the vast night of space behind him, is striking.

Apollo 17: Last Flag on the Moon

Apollo 17, carrying Commander Eugene Cernan, Command Module Pilot Ronald Evans, and Lunar Module Pilot Harrison Schmitt, touched down on the lunar surface on December 11, 1972. It was the sixth manned Moon landing and brought to twelve the number of humans who have walked on the Moon. It was also the final human spaceflight to the Moon by the Apollo Program or by anyone since. Gene Cernan had ridden there before on Apollo 10, but this time he was scheduled to get out and walk around, and perhaps even drive a bit. Harrison "Jack" Schmitt was making his first and only flight, but what a career opportunity it was for him. As a Moon-walking astronaut he was one of a kind. He had a Doctorate in geology from Harvard University and was one of six scientists selected by NASA as astronauts in 1965. All of them had been required to hold doctoral degrees in the sciences to qualify: three were physicists, two were physicians, and one, Schmitt, was a geologist. He was assigned to the crew of Apollo 17 to ensure that a professionally trained geologist was there to assay the lunar surface, and after Buzz Aldrin, Schmitt became only the second PhD to visit the Moon. Command Module Pilot Ron Evans also holds a first. While Cernan and Schmitt worked, slept, ate, and drove on the surface, Evans spent three days in orbit around the Moon. He is still the last person from Earth to have orbited the Moon alone.

Apollo 17 Crew: Harrison H. Schmitt (*left*), Ronald E. Evans (*right*), Eugene A. Cernan (*seated*), 1972. NASA Image.

Apollo 17 achieved a number of firsts. Harrison Schmitt applied his geology skills to sample material from the lunar highlands and to search for evidence of ancient volcanic activity. Apollo 17 had the longest stay the Moon (seventy-five hours), spent the most time exploring the lunar surface (twenty-two hours, four minutes), traveled the farthest in the Lunar Rover (21.75 miles), and brought back the largest quantity of Moon rocks (243 pounds).

Regrettably, Apollo 17 also represented several lasts. It was the last Apollo Moon mission, the last crewed spaceflight beyond Earth orbit, and the last time human beings set foot on the Moon. On December 14, 1972, Naval Captain Eugene Cernan climbed up the ladder of the Apollo 17 Lunar Module, worked his way into the cabin and sealed the hatch, and thus became the last human being to walk on the surface of the Moon. More than forty-six years have passed since then, but his "record" is in no danger of being broken.

Before they finished their final Moonwalk, as a salute to the Apollo Program and a reminder to others of where we came from and how far we could go, Cernan positioned his camera so that Schmitt and the American flag were framed in the black sky, with the flag pointing to the distant blue Earth. In the photograph Gene Cernan's reflection can be seen in the visor of Schmitt's helmet.

The American flag raised by the crew of Apollo 17 was unlike the five that preceded it. It was slightly larger, six feet wide rather than five, and it enjoyed a unique heritage; it had been to the Moon before. The crew of Apollo 11 had carried it to the Moon and back in 1969, so that it could be mounted in Houston's Mission Control Room in recognition of the support team's role in making the Moon landings possible. It had hung there until this flight. The crew of Apollo 17 returned it to the lunar surface to commemorate the entirety of the Apollo Program. It became the last flag on the Moon.

Skylab: The Forgotten Space Rescue

The Apollo 17 Command Module splashed down on December 19, 1972. I'd returned to Houston from my assignment at the Kennedy Space Center that September and been given responsibility for a new task titled Skylab Entry Support. Our team task was to develop the reentry monitoring and backup control plans for the Apollo Command Module reentry for the newly defined Skylab space missions, support the NASA team at the Mission Control Center during the flights, and perform analysis afterward to see how the CM onboard guidance programs had actually behaved during its reentry—all tasks we had done many times before. But while the Apollo missions had lasted a week and a half at most, the Skylab mission astronauts would be in orbit for months. It wasn't exactly the fast-paced work schedule of Apollo, so the team would have plenty of time to get it all done. That was good because our "team" had gotten much smaller; now it consisted of me and another aerospace engineer relatively new to the TRW crew, Jim Snowden, and he didn't need much leading.

Jim had a BS in physics and an MS in aerospace engineering with a specialty in flight and orbital mechanics—my kind of guy. He'd spent

Skylab manned orbiting workshop. NASA Image.

the prior twelve years at North American Aviation in California working on U.S. Air Force missile systems, until North American won the contract to build the Apollo Command and Service Modules. Jim switched over to engineering and testing the Apollo Command Module systems. As with many contractors, once these Moon machines were built and flying, the government funding dropped off, and North American Aviation had to let some people go. Jim was one of them. My transfer to Florida in November 1971 had created an opening at TRW in Houston for someone who could take over my reentry support responsibilities for Apollo 16 and 17. Jim applied for the position, and with his credentials, TRW hired him. He packed up and moved to Texas. That worked out well, since by the time I returned, my former boss, Bob

Manders, had made Jim part of the Apollo reentry support task team. He'd supported the crew training for and subsequent reentry of Apollo 16. NASA's Dave Heath had taken Jim under his wing, and now Jim was preparing for the launch and reentry of Apollo 17.

Just after I got back Dave invited me to sit in on a reentry briefing that he and Jim were to give to Gene Cernan, the commander of the upcoming flight of Apollo 17. Dave was also responsible for reentry on the Skylab missions in which I'd have a role, so I guess he wanted to remind me what the job was all about. I was glad it would be Jim and me working on Skylab and that Dave would be our Task Monitor. I'd been away from doing the technical work preparing for an upcoming spaceflight, and I missed it. The studies I'd been doing to get TRW ready for a possible role on the Space Shuttle were fun, but there's nothing like the rush of watching a flight you've been working on lift off the launch pad.

And what was Skylab?

The Skylab Program was a bit of a "bait and switch" move by NASA to allow them to survive the cancelation of the last three Apollo flights. In 1969 they contracted with the McDonnell Douglas Corporation to convert an existing Saturn IVB rocket stage, now going unused, into an orbital workshop—that is, into an early version of a manned space station (there were still no female astronauts at this time). It was a massive structure, 115 feet long and weighing over 200,000 pounds, and it was to be lifted into orbit atop an unused Saturn V rocket, the one that would have taken astronauts to the Moon had the Apollo 18 mission flown. Astronaut teams were to visit the space workshop in shifts, each spending months at a time in orbit performing experiments to discover how the human body responds to long duration spaceflight. NASA's original plan called for alternating a Skylab launch with an Apollo launch. Budget cutbacks killed that idea, and the first scheduled Skylab launch was postponed until May 1973.

In early concept drawings Skylab looked like an enormous farm silo with two large wings of solar cells sprouting from its sides. It had no rocket engine of its own, which reinforced its stovepipe appearance. A dish telescope, a solar observatory called the Apollo Telescope Mount (ATM), was mounted at a right angle to one end of the workshop. The

ATM had four solar-cell-paneled wings of its own, arrayed in a railroad-crossing X-pattern that dwarfed the winged cylinder beneath it. The design was a strange and unearthly looking beast, and the spaceship that McDonnell Douglas built looked just like those drawings.

Three astronauts at a time would travel to the workshop in an Apollo Command and Service Module (we had five of those left over from the canceled Moon missions), and a Lunar Module docking adapter and airlock would allow them to transfer from the CSM into the station. The living and working space inside was as roomy as a small house, a major improvement over the cramped and messy cockpit of the Command Module. Intended for astronaut stays of up to ninety days, the station had a shower, a toilet, and a ward room where astronauts could work and eat. Each could sleep in his own closet-sized bedroom. In addition to accommodating a sleeping bag and storage for his personal gear, the entrance to each astronaut's quarters had a curtain that could be drawn for privacy.

Between January 1973, when I started working on the program, and February 1974, when the third manned mission to Skylab splashed down, I had a great time. It turned out the Skylab Program gave me a chance to use my aerospace engineering training a bit more. As with Apollo, my duties on Skylab included those always intriguing "special studies." Though my engineering degrees had focused on high-speed atmospheric reentry, I was also reasonably proficient in "orbital dynamics," which translates to understanding how a spacecraft moves around in orbit and how to use its rockets to maneuver from one orbit into another. In fact, that's a concise definition of rocket science.

John Gensert was my manager at TRW at that time, and after Apollo 17 returned to Earth, Jim Snowden came over to work for him. John was a former Navy jet pilot who'd made numerous take-offs and landings from aircraft carriers during the Korean War, and he was one of the best managers I've had. John discovered that my background included orbital science, and he offered my talents in that area to NASA. I'm not sure whether it was Dave Heath or someone else, but NASA agreed to assign me to some tasks unrelated to Apollo Command Module reentry.

I had an opportunity to do two studies related to the orbit of the

Skylab workshop itself. I helped compute the range of paths traced across the Earth, the so-called operational trajectories, that the space laboratory would follow as it orbited the planet. Nailing those down was important, since the astronauts in the CSM would have to chase Skylab's orbit around the Earth to catch up and dock with it. I also prepared a "perigee altitude" report, where I calculated how close the workshop would come to brushing Earth's atmosphere at the lowest point in its orbit. Too close was bad, unless you were planning for Skylab to burn up in the atmosphere and come crashing down. Doing those two studies was invigorating; I hadn't flexed those particular engineering muscles since graduate school. I didn't get to work on the International Space Station later on, so it's rewarding now knowing I had done something useful for the flight of America's first space station.

Jim and I together developed the so-called splashdown footprints for the Skylab crew's Command Module reentry; that is, what area of the Pacific Ocean was available for landing the deorbiting Command Module. Jim also gave a presentation to Phil Shaffer, that once-feared audience of mine. Jim proposed a reentry flight option that would place the Command Module splashdown close to the Hawaiian Islands. When he finished, Phil surprised him by just saying, "Implement it." Not exactly the "Sit down" response I got when I briefed him on Apollo 15.

When these missions finally took flight, Jim and I were with Dave in the support room at the Mission Control Center during the Skylab CM reentries. We wore headsets so that we could follow the spacecraft-to-ground communications, while Dave communicated directly with the Flight Director in the Mission Control Room who was responsible for reentry. That part was energizing for us, but it turned out that the earlier launch of the Skylab laboratory itself was anything but amusing to NASA.

On May 14, 1973, a Saturn V rocket with the unmanned Skylab workshop mounted on top of it lifted off from Launch Complex 39A at the Kennedy Space Center. The flight immediately went wrong. During the turbulence of launch, the telemetry data transmitted from the spacecraft to Mission Control indicated that some components were malfunctioning, and internal temperatures were rising too fast. Once

John Aaron and Mission Control team supporting the damaged Skylab workshop, 1973. NASA Image.

Skylab settled into orbit, Houston flight controllers sent it a command to extend the solar panels wings automatically, but they wouldn't open. And the temperature inside the workshop was still rising rapidly, meaning that the solar shield intended to keep the spacecraft cool wasn't in place either. Unless someone figured out what to do, no one would be able to work inside Skylab. There was one piece of good news: the Apollo Telescope Mount, previously considered the most complicated piece of machinery to deploy, had opened up just fine. A set of automatic motors rotated it ninety degrees, locked it in place, and opened its four solar panels. If only the rest of the steps had gone that well. The first launch of astronaut visitors to the laboratory, scheduled for the next day, was delayed while NASA worked on what to do next.

Fortunately NASA's "failure is not an option" culture was still in play. As with the effort on Apollo 13, the engineers and astronauts on the ground spent the next ten days poring through their data to assess the extent of the damage. They determined that the thermal shield had been destroyed, and one of the solar panel wings seemed to have

Skylab 1 Crew (*left to right*): Joseph P. Kerwin, Charles "Pete" Conrad Jr., and Paul J. Weitz, 1973. NASA Image.

broken off. NASA teams around the country turned their attention to engineering and testing ways to work around the problems. Astronauts and technical teams rehearsed and reworked the gamut of actions the first crew might need to take to repair or replace the damaged parts they discovered when they arrived at the orbiting workshop, as well as to the list of tools and special equipment they should carry with them into orbit. When they had worked out several options, NASA made the decision to launch.

On May 25 Skylab 1, the first manned mission to Skylab, lifted off on a Saturn 1B rocket with Joe Kerwin, Paul Weitz, and Apollo 12 veteran Pete Conrad on board. Their first attempt at repair didn't work. Before docking, Conrad maneuvered the Command and Service Module around the Skylab workshop to inspect the damage and transmit video images to Mission Control.

"As you suspected, solar wing 2 is completely gone off the bird. Solar wing 1 is, in fact, partially deployed."

He also reported that the sun and micrometeorite shield was gone, and that a metal strap, possibly from the shield, was wrapped around the remaining solar panel wing, keeping it from opening. The crew took a break to eat and prepare for their first spacewalk (an Extra Vehicular Activity or EVA) to make repairs. Conrad moved the CSM close to the jammed solar panel and attempted a tricky EVA. The astronauts donned their spacesuits, and Conrad depressurized the Command Module cabin. Paul Weitz stood up in the open cockpit, and while Joe Kerwin held his legs, Weitz used a ten-foot "shepherd's crook" to try to pull the panel loose. He yanked on it several times, but it wouldn't budge.

"We pulled as hard as we could on the end of the [solar] panel," Weitz told Mission Control. "We couldn't get it out right now. . . . It's only a half-inch strip, but man, is it riveted on!"

Conrad saw that this approach wasn't working, plus he had to fight to keep the Command Module from colliding with the workshop. They needed to back off and talk to Mission Control about plan B, and there was no point in doing that while hovering outside. He resealed the Command Module and moved the CSM to the front end of the silo so that he could dock with it. That maneuver didn't go as planned either. Skylab had been designed to use the same docking probe as the Command and Service Module had used with the Lunar Module. But this time the probe wouldn't latch. After eight attempts, the crew decided to correct the problem manually. They suited up again, depressurized the Command Module cabin, opened the tunnel, and, just like home repairmen, took apart a section of the probe. They sealed things back up and tried again, and this time the latches worked. Now securely attached to the workshop, the crew took an extended break to get some sleep. It had been a long day.

When they were rested, they reopened the connecting tunnel and crawled through into the workshop to power it up. Paul Weitz entered first, wearing a gas mask until he could ensure that the internal air supply was circulating properly and to protect him if there were noxious gases present. Everything was fine, but it was hot in there, almost 130°F, so the astronauts had to work quickly on their next repair—installing a new thermal shield.

Astronaut Owen Garriott working outside the repaired Skylab workshop, 1973. NASA Image.

The crew had brought along a collapsible protective umbrella specially designed for that purpose; it was narrow enough to fit through one of the small airlocks in the hull that were intended for scientific instruments. The parasol was designed by Jack Kinzler, a NASA engineer whom insiders had nicknamed "Mr. Fix It" for his ability to devise inventive solutions to thorny problems. He built a prototype model using fiberglass fishing rods and parachute silk to demonstrate how his final design would work. NASA management approved it and directed his team to build a working version for the astronauts to carry with them to the workshop. Kinzler and his team had worked six days around the clock, without sleep, to build and test the parasol and have it ready for the launch of the Skylab 1 crew on May 25.

Pete Conrad and Paul Weitz slowly pushed the folded parasol through the small airlock and, using its built-in springs and telescoping aluminum tubes, opened it on the outside into a twenty-four-by-twenty-eight-foot aluminized umbrella. The jury-rigged sunshade worked just fine; the temperature inside the space station dropped immediately, eventually settling at 70°F. Jack Kinzler later received the NASA Distinguished Service Medal for his invention.

Two big problems solved, one more to go.

There was sufficient workshop battery power, plus a trickle from the jammed solar panel, to give them enough time to figure out how to unfurl the recalcitrant solar panel. They consulted daily with Mission Control, reviewing the pictures they'd taken of the solar wing during their first fly-around outside. On June 7 Conrad and Kerwin put on their spacesuits and went back outside. They tethered themselves to the hull, and Conrad extended a twenty-five-foot-long cutting tool, much like a gardener's pole pruner, in an attempt to sever the metal strap that had wrapped itself around the panel. Kerwin pulled on the lanyard that operated the cutting jaws a couple of times, but nothing happened. "Man, I am pulling," he reported. Conrad worked his way out along the pole by hand to see if the jaws had jammed. As he neared the tool, the jaws suddenly cut through the strap. The panel lurched partway open with enough force to send Conrad flying. Only his tether kept him from tumbling away into space.

Conrad pulled himself hand-over-hand back onto the hull, where he and Kerwin set to work getting the panel all the way open. They attached one end of a tether to the solar panel wing and the other to a support truss on the hull and took turns pulling on it, but the wing wouldn't budge. Conrad worked his way down the tether to the point on the hull where the wing was hinged to the spacecraft. He crouched down and slipped the tether line over his shoulder, stretching it out like a bowstring. Then he stood up and Kerwin yanked the other end of the tether again. A restraining bracket broke and the massive wing suddenly swung open all the way, this time sending both astronauts toppling off the spacecraft. Again the tethers kept them from slipping away for good; only a length of line had saved them from a certain death. Conrad and Kerwin regained their equilibrium and worked their way back to the hull. After making sure the solar panel was fully extended, they went back inside the workshop.

Full power was restored, the inside temperature was under control, and the uncooperative docking latch now worked. It was an amazing accomplishment demonstrating that humans could do demanding and meticulous work in the hazardous vacuum of space—a necessity if a future space station was to be occupied year-round.

Pete Conrad, Joe Kerwin, and Paul Weitz settled in for two weeks, setting up housekeeping, running experiments, and making observations of the Sun. On June 22, 1973, they undocked from the workshop, did a retro burn to reenter the Earth's atmosphere (using our updated monitoring and backup control procedures), and splashed down in the Pacific. The first house-call repair in outer space was a resounding success.

5

LIFTOFF MISSION TWO

SPACE SHUTTLE FIRST FLIGHTS

The Next Great Adventure

The NASA Space Shuttle Program operated for over thirty years, launching 135 missions between 1981 and 2011, but its origins go back another three decades. There have been many stories written about the triumphs and tragedies of the Space Shuttle, but not much about why it was built, or why it looked the way it did, or about those wizards behind the curtain—the people who made it all come together, some of whom remain largely unknown outside the specialized circles in which they worked.

From 1970 to 1984 at TRW, interleaved with my work on Apollo and Skylab, I had the chance to support some of the early planning for the Space Shuttle. And in 1984 I became a member of the IBM team responsible for designing, coding, testing, and supporting the Space Shuttle's onboard computer programs—the flight software that constituted the automated brain and nervous system of the Shuttle in flight. As with Apollo, my team there was one of many. The Space Shuttle Program was a vast, intricate, and complicated endeavor. Thousands of people were involved in making the Space Shuttle fly, including engineers, scientists, programmers, technicians, machinists, fabricators, researchers, managers, administrators, NASA and U.S. military personnel, and of course astronauts. My time on the Space Shuttle Program gave me answers to some of the whys and hows of the program, plus a chance to meet some of the men and women who made it happen. As I've attempted to do for Apollo, I hope to convey here a sense of the scope of this singular effort, and an appreciation for some of the unheralded people behind the scenes, a few of whom you might recognize from Apollo.

X-15: The First Spacecraft with Wings

I started working on the Space Shuttle Program before Apollo 17 lifted off for the Moon. In 1970 NASA was already in the early planning stages of the Space Shuttle Program—formally named the Space Transportation System—and TRW had won a few small contracts to help them with it. In April 1970 we received a study task from the NASA Manned Spacecraft Center to look at possible guidance systems for the Shuttle. Apollo 13 had just returned home safely and I had some time on my hands, so the assignment was passed on to me. I did some research and decided to take a look at another rocket plane that had flown into space in the 1950s.

The X-15 was an early experimental research program that tested the ability of a winged spacecraft to complete an atmospheric reentry from space, transition into flight like an airplane, and land horizontally on an airstrip. X-15 was an interesting beast: part jet aircraft and part rocket ship. It was designed and built in 1955 by North American Aviation (who would later go on to build the Apollo Command and Service Modules) under a contract issued jointly by NASA and the U.S. Air Force. Its only purpose was research; it was chartered to explore the design engineering challenges and flight characteristics encountered during hypersonic flight; that is, at velocities in excess of five times the speed of sound, or Mach 5. Because of the excessive stresses placed on the body of an aircraft flying at these speeds, learning how to design and build such a craft was essential if we were ever to fly our way into (and return from) space.

The aircraft looked like an elongated dart, with a cockpit, two stubby wings, a three-pronged tail, and a rocket engine added to the mix. Its airframe was the color of burnished coal, or of empty and airless space, the epitome of the description "jet black." It was capable of powered flight, but as it was a voracious fuel consumer, it was carried aloft attached to the underbelly of a B-52 cargo plane and set free at an altitude of 45,000 feet and a speed of 500 mph.

The X-15 made almost two hundred test flights between 1959 and 1968, and the list of its pilots reads like a Who's Who of spaceflight pioneers, everyone from Scott Crossfield to Neil Armstrong. It could soar to the edge of Earth's atmosphere, heights well above fifty miles at the very fringes of space, which it did thirteen times, earning those test pilots official "astronaut" wings. It still holds the world record for the fastest aircraft flight, a speed of 4,520 mph or Mach 6.7.

The X-15 was the first prototype rocket plane but not the only one. In the late 1950s and early 1960s other experimental so-called space plane designs were commissioned, including the U.S. Air Force X-20 built by Boeing and nicknamed "DynaSoar" (never flown), and NASA's HL10, for "Horizontal Landing," built by Northrop (thirty-seven flights). Even later, in the mid-1960s, the Air Force ran a series of classified design studies on launch and landing systems that could be either fully or partially reusable.

In January 1968, a full year and a half before Neil Armstrong's historic first step, NASA invited a number of aerospace specialists to present their ideas for a reusable spacecraft. Eighty professionals offered proposals, and while none of these became the blueprint for the Space Shuttle, many of the concepts did make it into the final design four years later. That same year, NASA officially began development work on the Space Shuttle by issuing a series of contracts to design a spacecraft that could deliver a payload to orbit, reenter the atmosphere, and fly back to Earth and land.

At TRW in 1970 I was interested in the X-15 because it was a human-operated, airplane-shaped reentry craft that transitioned to an unpowered flight and landing. It even had a series of small attitude control rockets on its nose and wings, which allowed it to maneuver at high altitudes where the air was too thin for traditional airplane flight. Part of the technology in the cockpit was the first "inertial navigation system," which gave the pilot the aircraft's position at the beginning of reentry and the direction to the landing site with unprecedented accuracy for that time.

Using a combination of the NASA specifications for the new design of the Space Shuttle Orbiter (the part that looks like an airplane) and the onboard navigation equations for the X-15 from a technical journal, I developed a simple software program to simulate the reentry guidance of a Space Shuttle using the X-15 navigation code. I made a set of computer runs to determine the series of roll and pitch maneuvers the Orbiter would execute during reentry to achieve a variety of cross-range and down-range distances. I thought the whole exercise was pretty cool. I turned in my results to the NASA people, who seemed pleased enough with them, but I didn't hear anything more. Given the somewhat simplified approach I'd employed, my guess is that what I did was filed away and never looked at again. The exercise did give me insight into the Space Shuttle's reentry flight though, which came in handy later on.

The X-15 test program itself contributed a great deal of information useful for both the Apollo and Space Shuttle programs, including training two future astronauts, Neil Armstrong and Joe Engle. To quote from NASA's website, the X-15 "provided an enormous wealth of data on hypersonic air flow, aerodynamic heating, control and stability at hypersonic speeds, reaction controls for flight above the atmosphere, piloting techniques for reentry, human factors, and flight instrumentation. The highly successful program contributed to the development of the Mercury, Gemini, and Apollo piloted spaceflight programs as well as the Space Shuttle."

Launch the Shuttle from North Carolina?

By the fall of 1971 my TRW work for the Apollo flights had really slowed down. Apollo 15 splashed down that August, and the next launch, Apollo 16, wasn't scheduled until April the following year. But the Manned Spacecraft Center in Houston wasn't the only NASA site gearing up for the Shuttle. In November 1971 I was transferred to the

TRW offices in Cape Canaveral, Florida, to work on a Space Shuttle launch planning task that TRW had received from the NASA team at the Kennedy Space Center. Barbara and our two-year-old son Paul made the trip with me. We rented a two-bedroom house on a corner lot with a grassy yard, front and back, in nearby Merritt Island. It was spacious and private, our first "real" house.

I was one of two TRW engineers given the assignment to work with NASA. The other was Bill Lee, a fellow TRW transplant whose job had been moved to Cape Canaveral the year before. Bill and I were assigned to evaluate alternative launch sites for the Shuttle. Congress had instructed NASA to consider locations other than Florida for the new Space Shuttle. A substantial windfall in dollars and jobs would flow into the locality chosen, and there was considerable political pressure to spread that wealth around. The task assigned us by KSC called for only two engineers, which was probably an indication of how seriously that center was taking the challenge. It became clear that it fell to Bill and me to help the NASA Kennedy Space Center brass keep Cape Canaveral as the primary launch site.

Bill was an interesting man. He was in his early thirties at the time, a smart, good-natured and exceptionally levelheaded coworker with little taste for organizational intrigue. He also had a sharp wit and dry sense of humor that a listener risked missing if not sufficiently attentive. His interests were astronomy and fast cars. In July 1972 he took his brand-new Datsun 240Z sports coupe, packed with a few clothes and his six-inch Meade refractor telescope, and drove to Nova Scotia, alone, to view a total eclipse of the Sun. He was a challenging, intelligent, and thoughtful guy to work with.

The Space Shuttle had some unique launch requirements as compared to Apollo, the most significant of which was the need to launch both eastward and southward; more on that later. Eastward launches were routine by that point, but southward launches could be a problem because of potential range safety issues. In short, if a launch was aborted and the spacecraft destroyed at some point following liftoff, the sizable and fiery remains might rain down like meteors onto a U.S. city, like Miami, or onto an island nation, like Cuba. To avoid that, a number of "no-launch" slices had to be carved away from the range of possible directions. The question became: did the direction of launch in the ones that remained allow the Shuttle to carry out its mission?

Cape Canaveral juts into the Atlantic and was capable of southerly launches, but it wasn't ideal. We evaluated several other sites, including Houston, Texas (not great east or south, and politically toxic since they already had the Manned Spacecraft Center); Vandenberg Air Force Base in California (terrific for southerly launches but pretty much a no-go eastward); and Cape Hatteras, North Carolina. That last location was surprisingly good, though it did lose an extra eastward "push" from the Earth's rotation by being located farther away from the equator than was Cape Canaveral. But all in all, it was an alternative that earned a closer look. Bill and I reviewed the pluses and minuses of possible sites and each played devil's advocate for the candidates.

That turned into an interesting exercise, since on several occasions when we were having lunch with colleagues in the cafeteria, I'd watched Bill, in his low-key and agreeable way, challenge both sides of a disagreement that arose, whether the subject was politics or religion or science or pop culture. After one of these sessions, I asked him if he had his own view on the topic, because I couldn't figure it out by listening to the questions he'd asked the others. Bill said he did, but instead of offering his opinion, he'd rather see how other people defended theirs.

Bill and I did come to agreement about proposing Cape Hatteras as a possible launch site for the Shuttle, and we presented our findings to the local NASA folks. They duly evaluated them and ran them up the line, and when all factors were included, in the end NASA and the Air Force decided the only feasible approach was to go with—surprise—two launch sites: Cape Canaveral for eastward launches and Vandenberg AFB for southward. Which had probably been their plan all along.

Some sources record the official beginning of the Space Shuttle Program as January 5, 1972, when President Richard Nixon directed NASA to proceed with the next generation of space transportation. As described in the president's speech, "This system will center on a space vehicle that can shuttle repeatedly from Earth to orbit and back. It will revolutionize transportation into near space by routinizing it. It will take the astronomical costs out of astronautics. In short, it will go a long way toward delivering the rich benefits of practical space utilization and the valuable spinoffs from space efforts into the daily lives of Americans and all people."

But the real planning for a space shuttle went back to the 1950s. And as I mentioned, NASA had officially kicked off a "Space Shuttle" program in 1968. The NASA Kennedy Space Center management had plenty of time to defend to NASA Headquarters the "engineering assessment" that Bill Lee and I did for them in 1972, so they could ensure that Cape Canaveral stayed at least a part of the action. History shows just how successful they were at that.

When my ten-month assignment was over, during which I'd gotten to view the launch of Apollo 16, I was transferred back to Houston, where I was given the job of supporting NASA's Skylab launches. Bill Lee stayed in Florida, and we crossed paths a couple of times after that. By 1974 we'd lost touch.

TRW Loses a "Can't Lose" Contract

Between 1970 and 1974 TRW had more employees than just Bill Lee and me working to get them ready for the Space Shuttle. We had other Space Shuttle study tasks paid for by NASA in addition to the one that precipitated my sojourn at the Kennedy Space Center. There was also work paid for by TRW itself to get the company team educated and prepared for a follow-on contract. To everyone's relief, in mid-1973 NASA informed TRW that they had decided to award a Space Shuttle support contract to us on a sole-source basis; that is, they wouldn't require us to compete for the work against other interested companies. The government had justified that by defining the type of work we'd be doing for them—the scope of work, in government-speak—as being essentially the same as what we had done for Apollo, with the only major difference being the design of the spacecraft involved. That wasn't as big a stretch as it sounds: we'd used our Apollo skills on both Skylab and the later joint United States–USSR spaceflight called Apollo-Soyuz, both of which had designs that were at least somewhat different from Apollo's.

Our team was delighted. We'd have the chance to work together on a project we loved for at least another three or four years. The day we heard the news we took off early and went to the Pizza Palace to celebrate.

If you've watched the TV game show *Wheel of Fortune* you'll recognize something called the Mystery Round. If a contestant's spin lands on that wedge and the player then guesses a correct letter, he or she has the choice of either taking the face-up value of $1,000 or picking up the wedge and turning it over. The hidden side will read either $10,000 or "BANKRUPT." When we watch *Wheel of Fortune*, I usually shout at the TV, "Don't pick it up! How stupid are you? You've got $1,000 guaranteed and you're going to risk having zilch on a fifty-fifty chance?"

My bosses at TRW didn't play *Wheel of Fortune* very well. Upper management went back to NASA with a counter-offer that would add more tasks to our new contract, ones that would have us doing work for a broader set of groups within the Center. NASA accepted our idea, but since the additional work we'd requested was well beyond the scope of what we had been doing, they told us they'd need to change our upcoming contract from sole-source to competitive. That change meant exactly what it sounds like. This is where I shout, "How stupid are you?"

TRW's management agreed to the change.

In December 1974 NASA formally announced the new contract as competitive and in addition to us, McDonnell Douglas—another aerospace contractor with Apollo experience—chose to bid on it. Many of our company engineers and technical people were tied up working on our current contract and couldn't give the proposal their full attention. Still, when we submitted our "best and final" proposal offer to NASA, TRW management was feeling pretty good about it. I even got a certificate thanking me for contributing to "Our excellent proposal for NASA JSC Space Shuttle Engineering and Operation Support." (The Manned Spacecraft Center had been renamed the Johnson Space Center, or JSC, in 1973.) When the winner was announced in the first week of April, we discovered that "excellent" wasn't how NASA viewed it. We lost, and we lost badly. Not only were we not the lowest cost bidder, which was a bit of a surprise, but we also had a lower technical score than McDonnell Douglas, which had never done this particular kind of support work for NASA before. That outcome meant our team had written a lousy proposal. NASA's rules require them to judge you on what you write, not on some private sense of which company might be better qualified.

We had turned over the mystery wedge and it said BANKRUPT. Over the next two months eighty TRW employees lost their jobs. A few of them were transferred to a TRW Systems organization in Redondo Beach, California, to do Department of Defense work. I was offered that opportunity, but I declined. I'd worked on a military contract at GE and had left it behind to join the manned space program; I didn't intend to go back. A number of people were hired by McDonnell Douglas, the winner of the contract we lost, but I resented joining that company just to teach them what we already knew. Others of our team had to find work elsewhere; many of the technical people were hired by the numerous big oil companies located in Houston. My Apollo teammate Phil Moseley had gotten a two-year head start on us there.

Fortunately, I was handed another option. I was among fifty engineers who applied for work just down the street with IBM, which had recently won a contract to build the onboard and ground-based software programs for the Space Shuttle. All fifty of us were hired. On June 3, 1974, I started my new job at IBM. Without any new missions to support, TRW Systems Group in Houston shuttered their doors.

They say that changing your job is one of life's most stressful moments. That was true for me in 1974, starting in January, when we discovered TRW would have to compete for more NASA work, and in June, when I was hired by IBM. But my transition was eased by a more agreeable distraction in my personal life. On March 14, 1974, my second son Dave was born. I had some concerns about which company's insurance would pay for additional post-natal care if required, but that was readily arranged by IBM. Before Dave came along, I remember wondering how I could possibly love a second child as much as I loved Paul, and as with many of life's lessons, the experience answered itself. I discovered what countless parents before me had learned: love isn't a pie, where one piece consumed means there's less for others. Rather it's like the sea, vast and embracing, it expands to fill all the spaces it encounters. Paul is a child of Apollo, a fiery trailblazer, born as Apollo 12 returned safely to Earth. Dave is a child of the Space Shuttle, rising on a spaceship with wings.

A final snapshot, and a long overdue thank you that dates from the time before I left TRW to start work for IBM, is an anecdote that also illustrates the blind spot of my assumptions. When I first moved to Houston in 1968 and went to work for TRW Systems Group, I discovered that then, as now, employees wore ID badges to gain admission to the office building. There was a receptionist stationed at a desk just inside the front door whose duty was to ensure that only badged employees came through. The receptionist at TRW's desk was a very attractive woman, probably in her early twenties, who wore her miniskirt to great effect. Whenever I arrived for work in the morning, it seemed that to show her my badge, I always had to step around one or more of my male coworkers hovering around this young lady, whose name I now can't remember.

One morning when I came into work, she was seated alone at her desk reading a thin magazine. She looked up when I entered, recognized me and smiled, and I took the opportunity to chat with her. I asked what she was reading. She held up the magazine and changed my life.

It was a slick publication with a multicolored cover that included maybe a close up of the Moon or a biologist's drawing of a virus—I don't remember for certain. It was called *Science News*. I'd never heard of it. She told me it came out weekly and offered to let me page through it. I did, and I experienced a tectonic worldview shift.

Science News gave readers a cogent, well-written summary of each of the important science discoveries of the previous seven days, and it did so without an equation in sight. It was clearly intended for people curious about science but without the discipline-specific understanding of the discovery being reviewed. I handed it back to her and realized I had been assuming she probably worked there just to pay for weekends partying with her friends. I was profoundly wrong, and I've tried not to make that mistake again. I subscribed to *Science News* later that week and I've been reading it faithfully since. As it did then, it never fails now to surprise me. Though I've lost track of her, I'll never forget the debt I owe her for introducing me to *Science News* and for exposing me as the one who was shallow.

Female Technical Specialists in the Space Program

One of the most significant transformations I encountered in moving on from the Apollo Program era was a dramatic increase in the number of women engineers and technical professionals I worked beside on the Space Shuttle Program. Lonnie Marshall, an African American engineer, was my first manager after I graduated from Florida, and as I've said, he was one of the most accomplished leaders for whom I've had the opportunity to work. For the remainder of my career, though, encountering black men who held management or senior technical positions in the aerospace industry was a rarity. However, beginning with that first job at General Electric in 1967, and continuing until I moved on from space program work in 1984, there were always women among the technical professionals I worked with: one or two in my fifteen months at GE, more than twice that during my six years at TRW, and the number soared when I joined IBM in 1974. For many of these women, it was a degree in mathematics that provided an entrée into positions traditionally held by men.

It seems that World War II was a stimulus for an earlier generation of women to seek technical degrees, just as JFK was the inspiration for mine. When the United States entered the war in 1941, an acute shortage of men to do skilled work prompted President Roosevelt to order the U.S. government to hire more African American workers to fill the gap. By the late 1950s President Roosevelt's order had encouraged some African American women to earn a degree, and as depicted in the 2016 film *Hidden Figures*, NASA Langley Research Center hired a number of college-educated black women with backgrounds in mathematics to work out the complex technical calculations required to accomplish a spaceflight. Other NASA centers did likewise.

Then computers were introduced in the early 1960s, and NASA and its contractors required skilled people who could create software programs to run on them. Computer programming, which in turn was dependent on knowledge of advanced mathematics, was not widely taught to the men who graduated with engineering degrees, including to me as late as 1967. But these technically skilled African American women were ready and stepped in.

Higher pay and increased status encouraged more and more women of all colors to seek a college degree, and by the time the Apollo and Space Shuttle programs came along—with their even more sophisticated computers, and with these, a demand for even more "human computers" to write the programs—there was a rapidly growing cadre of skilled college-educated women eager to take on the task.

Poppy Northcutt was a young woman with a degree in mathematics who had joined TRW in a technical support role in 1965, four years before I was hired there. By the time I arrived she'd worked her way up to being "a member of the technical staff," a TRW job position on a par with the male engineers in the group. Poppy was the lead technical person responsible for developing the software code that could simulate (that is, mathematically calculate) the possible trajectories that an Apollo spacecraft might follow during its days-long return to Earth from the Moon. The results of her analysis fed directly into my own, which were designed to take over once the spacecraft reentered the atmosphere.

NASA relied heavily on Poppy's professional skill and insight. During the flight of Apollo 8—while I was at midnight mass learning space jargon from a parish priest—Poppy was in a Mission Control support room assisting her NASA counterpart in ensuring that the Apollo crew returned safely from the Moon. I don't know of any other NASA or contractor woman at that time who was asked to provide technical support in the Mission Control Center during a flight.

Poppy was also an early advocate of women's rights, which I first discovered by chance. I once held a door open for her, but Poppy said, "I can open it myself" and waved me to go ahead of her. She didn't sermonize; she was an activist. Later, and with TRW's blessing, she was appointed as the women's advocate for the mayor of Houston. She eventually left TRW to earn a degree in law and became a full-time attorney.

Later, when I was hired by IBM in 1974 to work on the Space Shuttle, I found that the women in technical staff positions there outnumbered the men. The Shuttle onboard software development and test organization I joined was made up of about a hundred technical specialists and engineers, and fifty to sixty of those were female software programming professionals and managers.

"The Space Transportation System"

The engineering assignment I was given at IBM was unlike anything I'd done for TRW on Apollo or anywhere else. To describe the challenges of that job, and the incredible people I worked with and learned from because of it, I'll describe how very different, in both objectives and design, the Space Shuttle was from Apollo.

Though NASA had issued some early design contracts in 1968, President Nixon's directive in 1972 set requirements for this state-of-the-art breed of spacecraft that went well beyond just getting into orbit and flying back to land. NASA was told that this new program must include a reusable launch and landing vehicle; it must support a future space station; it must place commercial satellites into orbit; it must conduct scientific studies; it must meet classified mission requirements imposed by the U.S. Air Force; and it must substantially reduce the cost of manned spaceflight to $10 million per launch (in 1972 dollars). Though NASA ultimately achieved many of these objectives, that low launch cost wasn't one of them.

The original plan called for twenty-four launches a year, with development and operational costs divided between NASA and the U.S. Air Force. That objective proved too optimistic by a factor of five, especially later on when the Air Force dropped all funding for its development. The Space Shuttle went into operation in 1981 and flew a total of 135 flights over the next thirty years. The launch cost was certainly lower than for Apollo, which was about $1.9 billion per launch in 1969 dollars, but it still ran to at least $450 million per mission, more than forty times higher than the original target (and some sources argue that the $450 million figure is vastly understated). But in spite of some hardly trivial shortcomings, the Space Shuttle achieved most of the operational objectives set for it. And it became the workhorse that serviced the construction of a sophisticated human-occupied research habitat in space.

As noted, the formal name of the Space Shuttle is the Space Transportation System. The origin of the informal name "space shuttle" goes back to 1972, when President Richard Nixon used the term in announcing the program. Though most people think of the Shuttle as

Early concept designs for the Space Shuttle. NASA Image.

the airplane-shaped spacecraft that the astronauts pilot to and from Earth orbit, that vehicle is actually called the Orbiter, and it is only one of three primary components that make up the Shuttle. The other two, the Solid Rocket Boosters and the External Tank, are elements of the complex propulsion system designed to get the Orbiter itself into orbit, a measure of the difficulty of launching human beings into space.

Human spaceflight requires a vastly greater demand for support than does an unmanned rocket. Beyond the payload carried aloft—a satellite, a scientific experiment, or an International Space Station module, for example—humans need food and water and air and toilet facilities, as well as provisions for emergency medical care, protection from harmful radiation, and shelter from micrometeorite strikes. For extended stays, they also need some means to exercise. Accommodating all of that, plus the humans themselves, means that a lot of weight must be lifted into orbit.

Extremely powerful rockets, consuming enormous supplies of fuel, are required to produce sufficient thrust to climb the steep gravity well from the Earth's surface to a height of 250 miles, the orbit of the International Space Station. The irony of this demand is that the rockets and fuel are themselves tremendously heavy, requiring even more thrust—and so it goes. NASA realized early on that short of an unprecedented revolution in rocket fuel development—some as-yet-undiscovered substance that could produce more than a pound of thrust to the spacecraft for each pound of fuel consumed—the only solution was to shed weight as you go. Starting with the Mercury-Atlas two-stage rocket that John Glenn rode into orbit in 1962, rocket fuel was partitioned into separable sections, each of which was discarded before the next stage was ignited. That way the spacecraft engines wouldn't have to lift the deadweight of the spent fuel casings.

A similar approach for the Space Shuttle resulted in the design of multiple expendable sources of fuel for a launch. To gain the maximum thrust during initial liftoff required three bell-shaped Main Engines located at the rear end of the Orbiter itself, plus an External Tank (ET) that carried the fuel for the Main Engines, and two Solid Rocket Booster engines (SRBs) that were mounted on either side of the External Tank.

The Shuttle Main Engines carried no fuel of their own; that was fed to them from the attached External Tank. The ET contained two internal fuel tanks, one holding liquid hydrogen and the other liquid oxygen. Their contents were delivered separately to the Shuttle Main Engines via a complex of pumps and pipes and valves. There they were combined and ignited, and then, like aircraft fuel, could be throttled, metered, and even shut down. The two SRBs attached to the External Tank burned solid fuel, a complex mixture of flammable compounds and stabilizers. They operated much like fireworks set off on the Fourth of July. Once lit they couldn't be turned off.

For a launch, the Shuttle Main Engines were started first. The Shuttle system was locked in place for an additional six seconds after the Main Engines ignited to give them time to build up thrust, and all systems were given a final check. Then the Solid Rocket Boosters were lit and their lock-down bolts exploded, and the Space Shuttle launched. The SRBs continued to burn until their fuel was spent. At that point they were separated from the Shuttle and dropped via parachute into the Atlantic Ocean, where they were retrieved and refitted for another launch. After the External Tank's fuel was expended, the Main Engines were shut down and the ET was also separated to freefall into the ocean. The External Tank was not recovered or reused.

Once the Space Shuttle Orbiter reached orbit it relied on its own set of liquid-fueled engines, the Orbital Maneuvering System (OMS), which comprises two smaller, self-contained rockets located at the rear of the Orbiter next to the Shuttle Main Engines. These engines were used to achieve or change an orbit, to maneuver the Orbiter while in space (for example, to rendezvous with another spacecraft), and to perform the reentry burn when it was time to return to Earth. Finally, as with the Apollo spacecraft, a series of smaller fine-tuning rockets, called Reaction Control Systems (RCS), were located fore and aft on the Orbiter and used for shorter, precise adjustments to the Orbiter's position while in space, and to maneuver into position when it was time to perform the deorbit burn.

Once the Orbiter completed its reentry burn, the OMS and the RCS engines were progressively shut down. The OMS was not designed to be used during atmospheric flight, so the Orbiter had no engines burning to propel it forward during its reentry. The RCS system was used to keep the Orbiter stable until full aerodynamic flight was achieved and the wings and tail surface controls could take over. For its entire return flight through the atmosphere and landing, the Orbiter was an unpowered glider.

Why the Shuttle Has Wings

Besides cost per launch, there was another big program requirement for the Space Shuttle that dictated its overall design, but for which it was never used. That had to do with the shape and size of the Orbiter's wings. As I mentioned earlier, one of the prerequisites for funding the Space Shuttle was the ability to meet certain classified mission requirements imposed by the U.S. Air Force. None of those requirements was more severe, or caused greater impact on the size, shape, weight, and cost of the Space Shuttle than the mandate to support a USAF launch into a polar orbit.

As I had discovered during my ten-month assignment at the Kennedy Space Center in 1972, range safety concerns dictated the use of Vandenberg Air Force Base in California for Air Force Shuttle missions requiring a launch into polar orbit. A polar orbit is exactly what it sounds like: a satellite or spacecraft placed into such an orbit that it will circle the globe at an angle of ninety degrees to the equator, following lines of longitude and passing over both the South and North Poles. Vandenberg Air Force Base is located northwest of Santa Barbara, California, on a dry rocky outcrop that juts into the Pacific Ocean. A spacecraft launched due south from there passes over nothing but a vast expanse of ocean all the way to the Antarctic Circle. In short, there are essentially no range safety issues associated with launching south into a polar orbit from Vandenberg.

The Air Force had several reasons for wanting to put a classified satellite into a polar orbit. These were all tied to the fact that as the satellite continues to circle from south to north, the Earth beneath it rotates from west to east, ensuring over time, the satellite's sensors will have access to every square mile of the globe, including all of the then USSR and China. However, the Air Force had an additional requirement for such a mission. They wanted to be able to launch the Shuttle and have it return to base in a single orbit. That way the Orbiter itself wouldn't pass over a populated and perhaps hostile landmass, particularly Russia or China, and have its presence or purpose discovered. One might speculate that the idea was to have the Shuttle carry a small undetectable classified payload as its cargo, open the cargo doors and release it as soon as they reached the altitude of the desired polar orbit, and then immediately deorbit and return to Vandenberg.

An object moving in a polar orbit 250 miles above the Earth, the approximate altitude now of the International Space Station, takes about ninety minutes to complete one orbit. In that same ninety minutes the Earth rotates east about 1,500 miles at the equator, or about 1,200 miles at the latitude of Vandenberg Air Force Base. That meant when the Space Shuttle completed one orbit, Vandenberg AFB was now 1,200 miles to its east, so once the Orbiter reentered the atmosphere and started its long unpowered flight home, it also had to fly 1,200 miles cross-range to get there. The ability to glide that far sideways is no small matter of design.

The Orbiter wings had to be wider and longer to provide sufficient lift and control, and that complex shape and size also meant the Shuttle couldn't use a simple heatshield ablative coating like Apollo, but rather needed a new protective layer made of a patchwork of individual heat-resistant tiles that completely covered the underside and front edges of the Orbiter. The Apollo heatshield, like the Command Module itself, was intended to be used only once. Its heatshield was engineered to melt away during reentry and to carry with it the tremendous heat built up around the spacecraft. The Space Shuttle, however, was designed to be reused, so its heatshield had be a different material, one that would protect the Orbiter during reentry and also be reused.

All of this also added weight, and that, plus the need to compensate for the loss of the speed boost that an easterly launch from Cape Canaveral would have provided, caused a significant redesign of the Shuttle's booster and engine thrust capability. In short, nearly everything about the Space Shuttle design we came to know was driven by the polar orbit requirement from the Air Force. The irony is that for a complex set of reasons, the Space Shuttle was never launched from Vandenberg Air Force Base. After years of delay, the first launch there was scheduled for Space Shuttle *Discovery* in July 1986. However, that January Space Shuttle *Challenger* exploded during liftoff from Cape Canaveral. The Vandenberg launch was first postponed and then canceled, and no future attempt was made.

Additional design modifications to the Shuttle as a result of the findings in the *Challenger* accident investigation made polar orbit launches using the Space Shuttle impractical, and the Air Force abandoned its plans for Shuttle use at Vandenberg AFB and shifted back to unmanned rockets there. Over the thirty-year life of the Space Shuttle Program, the Air Force did utilize the Space Shuttle for several non-polar orbit classified missions (estimated as fewer than a dozen) that were launched from the Kennedy Space Center.

The Onboard Computers

You can probably imagine that a vehicle as unique and revolutionary and complex as a winged spaceship with a human crew wasn't going to get by using computer technology as primitive as the Apollo Guidance Computer. The electronic "brain" controlling this new flying machine had to be built from scratch too. That need for innovation gave me the chance to work with the Space Shuttle team for the next ten years.

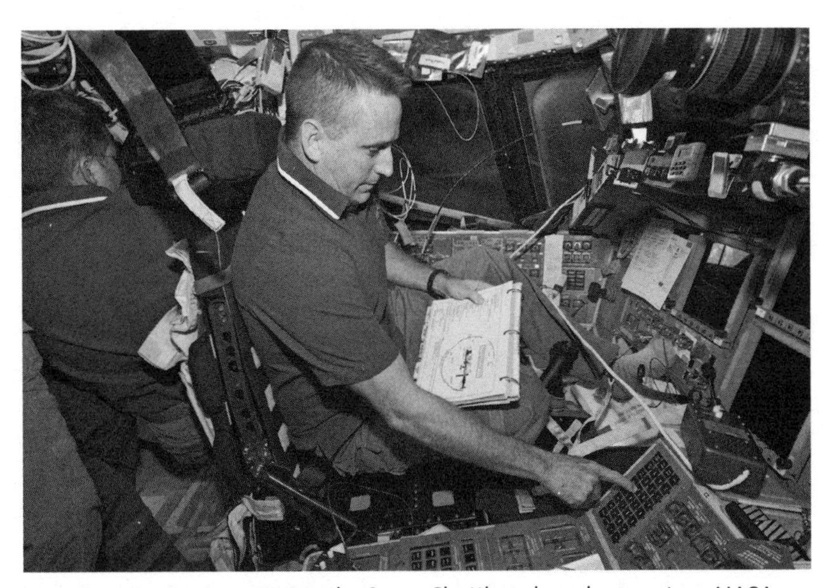

Astronaut Ken Ham operating the Space Shuttle onboard computers. NASA Image.

There was an enormous difference between the cockpit computers designed for the Apollo Command Module and those used on the Space Shuttle Orbiter a decade later. The Apollo Command Module Computer, circa 1966, had an astronaut interface that consisted of a calculator-like keypad, three five-digit numerical display registers, four simple function keys and a series of indicator lights. Its internal programs were literally hard-wired onto rectangular panels and couldn't be modified during a spaceflight. That meant that a dozen primitive integrated circuit boards, each designed to perform a specific in-flight procedure such as liftoff and ascent, star tracking, navigation, or reentry, were pre-inserted into the computer's file-drawer-like frame and manually activated by an astronaut as a mission operation required it. Only certain limited numerical data values and commands could be entered during flight. As I pointed out earlier, the Apollo Command Module Computer had very little memory, about half that of a Commodore 64 home computer.

In this technology the Space Shuttle was very different. The Orbiter's cockpit layout was similar to that of a jet fighter, befitting its X-15 heritage, and the onboard computer processor, called the flight computer, was built from a design developed in 1972 for an Air Force jet aircraft and previously used on Skylab in 1973. While a single computer processor weighed about sixty-four pounds, the Shuttle carried five of them onboard plus three hundred electronic black boxes and more than three hundred miles of wiring to interconnect them. The total onboard computer system network weighed over 17,000 pounds, more than the weight of the Apollo Command Module. The 1974 version of the Shuttle flight computer had the capacity to store and access about 100,000 words of data in its memory, primitive by today's standards, but it had three times the capacity of the Apollo computer and ran ten times as fast. In 1991 an upgrade more than doubled the Space Shuttle onboard computer's speed and memory.

As great an advance as this system was over Apollo, it was already outdated by the time the Space Shuttle launched in the 1980s. That pattern is a product of any complex state-of-the-art subsystems used in newly developed high-speed aircraft and spacecraft. The entire Space Shuttle was being designed from scratch, and a cardinal rule of such projects is that the so-called baseline, in this case, the computer hardware and software designs, must be "frozen"; that is, no design changes are permitted until the entire system is assembled, integrated, tested, and readied for operational use—a process that usually takes several years. Large and very complex systems are notoriously difficult to bring together into full operational performance, and changing individual subsystem design components as you go, with all the known and unforeseen consequential effects and redesigns on other dependent subsystems, is a recipe for disaster. Or at least for a large overrun in cost and a prohibitive slippage in schedule.

Not that this circumstance is unheard of in other industries. In my experience, most of the large government and commercial systems that get into trouble are affected by the "let's change it now to make it better" disease. The other affliction is "I don't have time to do all the tests now, we'll pick them up later during system test." In fact those two ailments usually occur during development of the same system.

Thanks to John Aaron (the same John Aaron of Apollo 12 "SCE to AUX" fame), the software programs that ran on the Space Shuttle's flight computers didn't suffer from too much design meddling or too little testing. "Changing it to make it better" wasn't simply a case of tinkering with success in the beginning; it was essential. These new programs, so unlike anything that had been attempted before, had first to be designed, and then completely redesigned, and finally redesigned yet again, just to get the Shuttle's flight systems to work at all. That effort consumed the entire IBM programming and engineering team in Houston, plus NASA and North American Rockwell in Downey, California, who were building the Orbiter, for the better part of two years before the onboard computer software could be approved for flight on the Shuttle.

Fly-by-Wire

The Space Shuttle Orbiter was a so-called fly-by-wire aircraft. That means the astronaut had no direct manual control of the flight mechanisms: all control commands, aileron adjustments, jet firings, switch positioning, and joystick movements initiated by the pilot were first passed through the onboard computer for processing. This might seem unreasonable to an outsider: that an astronaut's skilled judgment could be trumped by a program (and programmer's) design? But in fact no human being was capable of flying the Shuttle unassisted by a computer.

The Orbiter itself was dynamically unstable: during high-speed atmospheric flight—that is, during launch and reentry—the extreme forces buffeting the spacecraft produced abrupt and violent oscillations that, if left unattended, would cause it to spiral out of control. To prevent that, the flight surfaces and control jets were engaged in continuous corrections to the aircraft's roll, pitch, and yaw positions to keep it stable, and these commands were so rapid, so precise, and so frequent that they were beyond human response times. For much of the flight time in the atmosphere the onboard computer had sole control of the spacecraft, with the astronaut-pilot playing a monitor role. However, even during those periods when the human was free to take over, pilot commands were first evaluated and modified by the computer before being executed, to ensure they didn't inadvertently cause the Orbiter's motion to stall, spin, or become otherwise erratic.

The Space Shuttle was an early adopter of fly-by-wire technology but not the first. A Soviet Tupolev ANT-20 aircraft was outfitted with it in the 1930s, as were Canadian and British aircraft in the '50s and '60s. Commercial use came later: an Airbus A320 was the first airliner to use an all-digital computer-controlled, fly-by-wire system in 1984. However, in the early 1980s when the Space Shuttle flew, these advanced fly-by-wire systems were still under test and operated only in limited service. The Orbiter's system was the first to be put into use in an extreme flight environment. The Shuttle computer and its associated flight software programs had to be utterly reliable.

Computers Control Everything

The computer selected by NASA for use on the Space Shuttle was an IBM model AP-101 flight computer, a modified version of the IBM AP-1 used on the Air Force F-15 jet fighter. In March 1973, a year and three months before I started work at IBM's offices near the Johnson Space Center, NASA had awarded IBM's Houston group the contract to

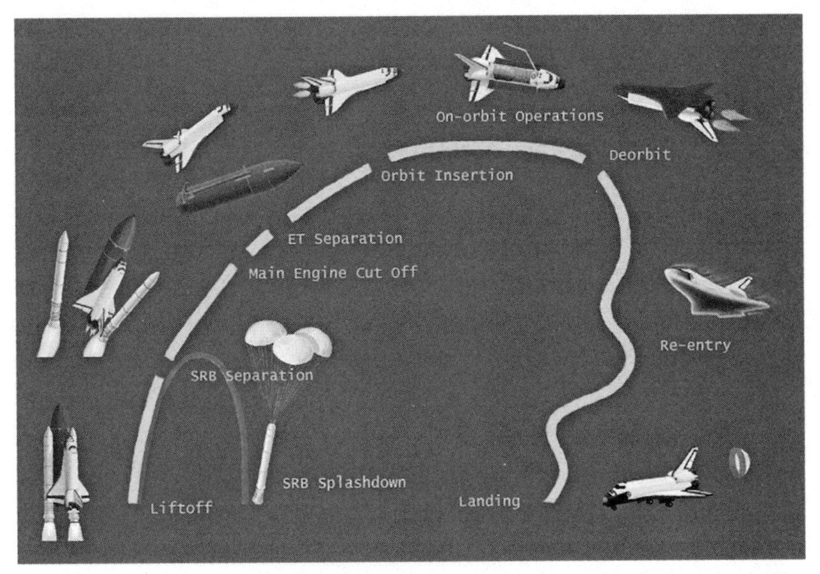

All phases of a Space Shuttle mission controlled by onboard software. NASA Image.

design, build, and test the entirely new software programs that were to run on this computer. Because the flight computer's performance was critical to both mission success and crew safety, NASA imposed a "fail-operational/fail-safe" mandate on the system, meaning that if a single system failed the mission could continue, and after a second failure, the mission would be aborted, but the crew must be able to return safely to Earth. This requirement in turn dictated the use of four identically programmed onboard computers that ran simultaneously on each flight, each one processing the same input information and issuing the same commands. There was also one additional backup computer running onboard that was held in reserve.

Each of the four primary computers "believed" it was in control of the Shuttle, but only one of them actually commanded the vehicle at any given time. The four computers continually compared results, and if a conflict arose, the three in agreement "voted" the conflicting unit out of the loop, shifted command to one of the remaining computers if necessary, and continued with the mission (fail operational). They also alerted the crew with warning lights, audio signals, and display messages that one of the computers had failed and been taken out of the set.

If one of the three remaining computers subsequently failed, the voting process was repeated and the two computers in agreement voted out the third one that differed. In theory the mission could continue with just those two, but if a third failure later occurred there would be no way to identify the incorrect one. So instead the mission was aborted and the spacecraft and crew would immediately return to Earth (fail safe). To reduce risk further, the fifth onboard computer, the one designated as a backup, could also come into play if called upon. This computer was independently programmed by another contractor, North American Rockwell, to ensure that the same software or programmer mistakes weren't likely to exist in all five computers. The backup also ran with a reduced "bare bones" code that could not complete an entire Shuttle mission but could stabilize the Shuttle and, with the assistance of active manual control by the astronauts, return it safely to Earth.

Talk about complicated. But in those days, that was how NASA insisted systems be designed if human lives were at risk. In practice the backup computer was never called upon to take command of a Space Shuttle during any mission.

When awarded this contract, the IBM Houston group was primarily a software development team. They quickly realized that NASA was an engineering organization that didn't understand programming (or programmers) very well, so they brought on some engineers of their own. That was where my new job came in. Barely a year after the last man walked on the Moon, NASA was about to roll out a radical design for a manned spacecraft in which nearly everything onboard would be controlled by computers, and I do mean everything. Furthermore, its computers would be running an entirely new and never previously flown set of software programs during its first operational use.

This new system would be called upon to control the launch from Cape Canaveral, including any abort procedures that might arise. It would perform all maneuvers while in space, to include achieving, maintaining, and changing the orbit, and carrying out all payload operations, which entailed the deployment and recovery of satellites. Finally, it would initiate and control the reentry flight back through the atmosphere, navigating to the landing site at either Kennedy Space

Center in Florida or Edwards Air Force Base in California, and putting the Orbiter down safely on the runway.

By comparison to the Space Shuttle, Apollo was a Model-T Ford. No set of computer-controlled spaceship operations like this had ever been attempted, and the computing system required to perform them demanded a software design fundamentally different and far more complex than the primitive system used for Apollo. Nothing that got us to the Moon could be reused here, and so it was discarded.

Engineering the Onboard Software

IBM had occupied offices across the street from the Johnson Space Center for several years; they'd been providing computing support for the NASA Mission Control Center since the early 1960s. But when they won the Shuttle contract they quickly understood this was no routine programming task; NASA and IBM had to start over. NASA had little expertise in the programming techniques that were to be employed to make this new type of computer operate, and it was equally true that the IBM programmers in Houston didn't fully understand the engineering complexities of what it would take to control and fly a spaceship like the Space Shuttle.

IBM brought in senior engineers and scientists to bridge the gap, to act as "translators" between the hardware engineers at North American Rockwell (and other contractor companies), who were building the complex innards of the Space Shuttle, and the IBM software developers, who were to design, develop, and write the code. Some IBM engineers with military aircraft experience were transferred to Houston from locations like Owego, New York. Others with Apollo manned spacecraft experience were hired locally. The fifty engineers and technical people, including me, who were hired by IBM after the TRW work ended, filled in the gaps. I joined a small group of a half-dozen engineers assigned to work on the front-end requirements (that is, figuring out exactly what NASA required the software to do), including one certified rocket scientist already wearing an IBM badge, a brilliant guy named Gary Smith, who'd earned his PhD in orbital mechanics (technically, astrodynamics) from UCLA.

Because I came on relatively late, all of what I thought of as the "cool" areas of Shuttle's onboard functions—guidance, navigation, and flight control—had already been assigned to others. Guidance, Navigation, and Flight Control, abbreviated GNC, are the holy trinity of onboard computer systems, whether in a spacecraft or in your automobile. Navigation keeps up with where you are at all times. Guidance figures out the trajectory, flight path, or set of roads you need to take from where you are currently to your final destination. Flight Control executes all the maneuvers needed to get you there and keeps you from crashing into things and other undesirable happenings along the way. In your car, your "navigation" system can do both guidance and navigation while you do the flight control (at least for now). On the Space Shuttle, the computer did all three. Two of IBM's young and talented aerospace engineers, Bill Madden and Rick Padinha, had transferred to Houston from other IBM locations. Bill's expertise was jet aircraft guidance, and Rick's was spacecraft flight control, so there wasn't much discussion about which of them got what. Of course, Gary had responsibility for all of the in-flight navigation, including orbital rendezvous and operations—an extremely complex and difficult area that he was supremely suited to tackle. Gary was in rarefied company; another notable rocket scientist, Apollo 11 Astronaut Buzz Aldrin, wrote his PhD dissertation at MIT on the subject of orbital mechanics.

So I was handed the "leftovers": working up the software to support the cockpit displays and controls. I wasn't too excited about it at first; it seemed to have little to do with engineering and more to do with the numbers and graphics and switches that went with the crew's cockpit gauges, readouts, and display screens. Two circumstances changed my mind. First, the displays and controls in the Orbiter cockpit provided the status of, and control over, nearly all the hardware and software systems on the Space Shuttle. So I'd need to get a detailed grasp of every element of hardware or subsystem that the onboard software touched, and that was pretty much all of them. Second, NASA required the astronaut crew members to understand thoroughly how everything on the Space Shuttle worked, and to be sure that the data displays and controls they were provided were the correct and useful ones. So they'd assigned engineering design of the Orbiter's cockpit displays and controls

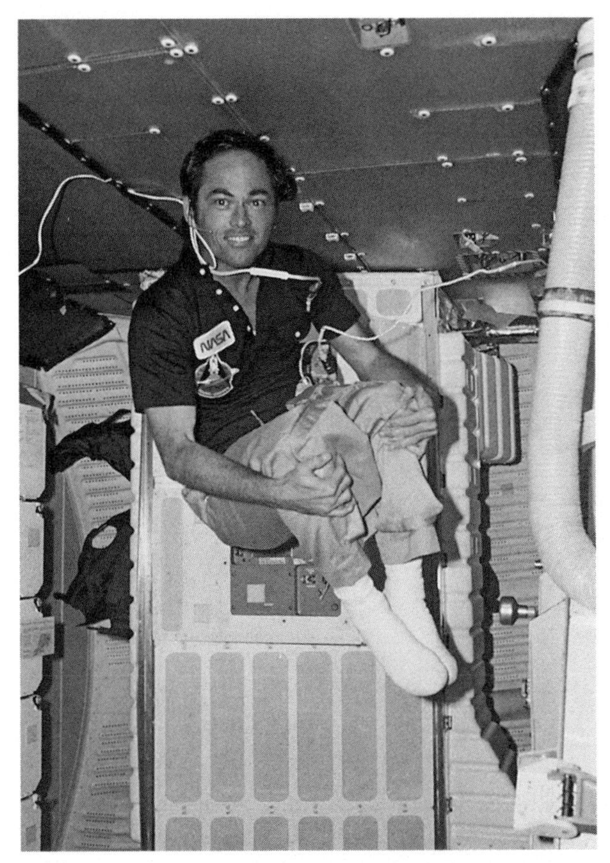

Astronaut Bob Crippen onboard *Columbia* on the first Space Shuttle flight, April 12–14, 1981. NASA Image.

to two newly selected NASA astronauts, Robert Crippen and Richard Truly. That meant while my colleagues were flying off to meetings with engineers in California or Massachusetts, I'd be strolling across the street to sit down with astronauts.

Working with Astronauts

In 1974, as today, the astronaut offices were located in Building 4 on the campus of the Lyndon B. Johnson Space Center in Houston. Though JSC was literally across the street from my office, the campus itself covered more than two and a half square miles, and Building 4 was on the far side. So I couldn't exactly stroll over. I drove. It wasn't my first visit to Building 4. I'd supported a couple of crew briefings there during my

Apollo days, and once I even rode in an elevator with Alan Shepard. We were headed to different meetings, and though he didn't smile at me, he did nod. I didn't have the nerve back then to strike up a conversation. I was accompanied by my NASA task manager, and I never strayed outside the conference room.

This time I'd been invited to a meeting by a secretary in the Astronaut Office who had contacted my boss at IBM to find out who was working on the Orbiter displays and controls software. I'd expected to find a briefing room filled with NASA staffers, but it turned out to be an ordinary two-person office, about the same size as the one I shared with Bill Madden at IBM. The door was open, and two people I didn't recognize were sitting at adjacent desks working. I knocked on the jamb, certain I was at the wrong place, but one of them stood up and walked over to greet me. He was only a few years older than me. He looked fit, and his face was as tanned as a lifeguard's; he didn't have the pasty complexion of many of the NASA folks I'd worked with previously.

"You must be the guy from IBM," he said.

He had a big, white-toothed grin. He was the handsome "Captain America" type, like Neil Armstrong and Gene Cernan, that I'd expected an astronaut to look like. I nodded and he shook my hand.

"Come on in. I'm Bob Crippen, this is Dick Truly," he said, pointing to his office mate.

Truly looked up from whatever he was doing, took off his reading glasses, and waved me in. He was about the same age as Crippen, but his face looked more scholarly than chiseled, especially wearing those glasses. He motioned me to a small conference table and moved over to it himself. I sat down opposite him, and Crippen dragged his chair over.

"So is IBM going to give us the displays we need?" Truly asked with a smile.

Ah, so he was an astronaut too. Wow, and these were the guys I'd be working with. That was how it started.

Bob Crippen and Dick Truly had a lot in common. Both were aerospace engineers, both were Navy fighter pilots and test pilots, and both had been assigned as astronauts to the Manned Orbiting Laboratory. MOL was a short-lived U.S. Air Force predecessor to Skylab, designed to put a human-occupied space station into low Earth orbit. Though the mission was classified at the time, its purpose was to allow high definition orbital surveillance of potential military threats. The program was announced in 1963 and canceled after its budget was cut in 1968, with flight tests under way and seventeen astronauts selected and in training. Seven of the MOL astronauts, including Bob Crippen and Dick Truly, transferred over to NASA.

I worked with Crippen and Truly for the better part of a year. I discovered they had a deep understanding of the engineering that went into the Space Shuttle and could dive into the detailed inner workings of both software and hardware, both the "bytes" and the "bolts." As I said, using the Orbiter cockpit's displays and controls, an astronaut could monitor, regulate, or operate virtually every element of the Space Shuttle system. These two men certainly knew their job. Working beside them to help the IBM team design the software was exhilarating.

As we got further along in the process, we all had to fly out to the North American Rockwell plant outside Los Angeles, California, to meet with the hardware engineers who were building the Orbiter. Crippen and Truly flew in their T-38 jets; I lumbered out in coach class on American Airlines. I rotated from meeting to meeting with them—there was a lot of hardware involved—and took notes as they asked the Rockwell engineers questions about the details of some obscure component of a control actuator. Then they'd turn to me with an equally pointed question about how the IBM software would respond to it. Hearing about the near-legendary competence of NASA astronauts was one thing; seeing it in action was humbling. They were two of the most impressive professionals I've ever met.

As I became comfortable with the relationship, I grew bold enough to ask them to autograph a Space Shuttle concept drawing I'd brought along to a meeting with them. At that point they didn't yet have official NASA photos of themselves.

Bob laughed at me. "Why in the world do you want that? We haven't done anything."

Astronaut Dick Truly onboard *Columbia* on the second Space Shuttle flight, November 12–14, 1981. NASA Image.

"Maybe someday you'll be famous and your signature will be worth something." I was halfway kidding then, but my prediction was on target.

Bob Crippen went on to fly four Space Shuttle missions, including the first orbital flight, STS-1, and he spent more than 560 hours in space. He later became the Director of the Kennedy Space Center in Florida and was awarded the Congressional Space Medal of Honor by President George W. Bush.

Dick Truly flew five Space Shuttle missions, including three of the Approach and Landing Tests flights and the second orbital flight, STS-2. He was later appointed NASA Administrator. Truly was awarded the President's Citizen Medal by Ronald Reagan. Bob Crippen and Dick Truly did sign that Shuttle drawing for me. I still have it.

Ten Pounds in a Five-Pound Sack

Getting the Space Shuttle software program requirements defined and documented was just the front end of the problem. Though it's hard to comprehend today, in the early 1970s computer hardware was a scarce and expensive resource, and labor was cheap by comparison. Like every government software program written in those days, or since, what the customer or user wants that code to do is always more than the programming team believes it can provide at the expected cost and schedule. The Space Shuttle software was no exception.

When my engineering colleagues and I conveyed all of the numerous detailed requirements demanded of the computer by NASA and the other contractors, and the IBM programmers coded them, the resulting software had two serious problems. One, it couldn't all fit in the memory space of the computer, and two, the portion that did fit ran way too sluggishly. So here we were, a couple of years into the IBM contract and basically nothing worked.

All hell broke loose. NASA called IBM senior management on the carpet and demanded to know what we were going to do about it. IBM, of course, blamed the other contractors—and remained judiciously silent on those requirements that came from NASA—for setting unrealistically high expectations on the limited computer resources available. The contractors in turn claimed that IBM didn't know how to write efficient code. Unfortunately both parties were right, which led to numerous finger pointing sessions, complete with abundant and detailed charts, that NASA somehow had to adjudicate.

When the smoke finally cleared, or the chalk dust settled, NASA wasn't convinced that we'd given it our best and said they wouldn't make demands of any other contractors to adjust their requirements until we produced the tightest, leanest, fastest code we could. So we engineers worked hand-in-glove with our software programmer colleagues for several months, "scrubbing" the code. Though the software had been written in a high-level language, called HAL/S by its inventor, Intermetrics Corporation, in Cambridge, Massachusetts, we had to examine the underlying assembler language code elements—that is, the arcane computer instructions lurking below the English-like programming code—to see where efficiencies might lie. As an aside, Intermetrics claimed that HAL/S stood for "High-order Assembly Language/Shuttle" but we believed it was actually a nod to the deranged computer HAL 9000 in the movie *2001: A Space Odyssey*, which in turn was rumored to be the evil twin of IBM, as in one set of initials rearward from "IBM." (Get it?)

I mentioned Gary Smith earlier, the rocket scientist who was responsible for the onboard navigation software. Gary led one of the IBM teams assigned to scrub this version of the flight software so that it would fit in the onboard computer, a tedious and often thankless job that earned him the mock superhero sobriquet "Compiler Man" from his coworkers.

In the middle of all this, Gary also introduced me to the "fandom" side of science fiction, which I later embraced. Gary was responsible for my first attendance at a World Science Fiction convention in Kansas City in 1976. That's where I encountered a young and then unknown actor named Mark Hamill. He was excited to be attending the gathering as a cast member of an upcoming sci-fi film with the clichéd title *Star Wars.* Attendees sat through a presentation about the film by Hamill and Producer Gary Kurtz and gave it a chilly response. Their reception at the 1977 World Science Fiction convention in Miami was very different.

Mark Luna was a computer specialist who worked with Gary Smith on the software scrub. And because Mark worked closely with Gary, he earned the title "Bit Boy" as part of the dynamic duo "Compiler Man and Bit Boy."

Mark, a Texas-born Mexican American, was a computer programmer analyst and still the best at that job I've met. At one point, I watched him respond to an angry phone call from a North American Rockwell hardware technician in California, 1,500 miles from our offices in Houston. The tech was livid because a bug in our onboard software code had screwed up his integration tests and all he had to work with was a nearly incomprehensible stream of printed data from his computer. Mark opened his desk drawer, pulled out his accordion-folded paper printout of the code, and calmly talked the tech through a detailed debugging of the problem. Remember, these were the days of voice-only telephone conversations and no access to electronic data transfer. Mark's intimate knowledge of the flight software allowed him to work with only a line-by-line transcription of the data, over the phone, by a fuming user in California, and figure out what was wrong and how to fix it.

Frank Donaghe was another outstanding computer software professional and a reason we got through the code scrubbing challenge successfully. I'd met Frank during my first year on the IBM team when he game-mastered a table-top role playing game called Space Invaders. (Yes, we had such games back in the Stone Age, and yes, we played them when we were supposed to be on the job.) During the "Great Onboard Flight Software Scrub" Frank became so adept at analysis, debugging, and problem resolution that one wag posted a sign above his desk reading "Keep This Area Free of Kryptonite." Frank and I have remained life-long friends.

In those next six months I learned more about the size and speed of "LOADs" and "MULTIPLYs" and "GOTOs" than I would ever need or want to know again. When we got it all done, we had substantially reduced the volume of code required and significantly increased the program execution speed. But even with all our efforts, the programs we'd built were still too big and too slow. Frustrated, but now at least a bit mollified, NASA attacked the remaining problem the way they always did: they put all of us contractors in rooms, with themselves in charge, and didn't let us leave until we fixed it.

Phil Shaffer Again

The room I was assigned to was half a continent away. As I said, cockpit displays and controls interacted with virtually every piece of hardware on the Shuttle Orbiter, and all of that was designed by North American Rockwell in California. Our job was like trying to fit ten pounds of potatoes into a five-pound sack. It wasn't going to be easy. Bob Crippen and Dick Truly were out of the picture for this round; they had more important tasks now, like training to fly the Shuttle. This time I'd need to satisfy Phil Shaffer again.

I'd tried that once before with my excess of charts, and it didn't go well. By 1976 NASA had given Phil, an Apollo veteran, the title of Manager of Shuttle Program Flight Techniques and Software Integration.

Me describing how the onboard software operates, 1983. J. Clemons Photo Copyrighted 2016. Penton Media. 123443:0516SH.

That meant he was in charge of having IBM and North American Rockwell resolve the misconnections between the requirements placed on the Shuttle's flight software programs and the capacity of that software to execute them. Phil was more than up to the task; he'd served as both Flight Dynamics Officer and Flight Director during the Apollo and Skylab Programs. Now he was flying to Downey, California, to run the "play nice together" sessions between IBM and North American Rockwell. That meant I'd be in a lot of meetings with him.

I wasn't looking forward to it. When I'd met him five years earlier during the Apollo Program, the encounter with Phil subjected me to what I would describe today as his version of *Shark Tank*. Now I was with a new company and had more experience dealing with the NASA "way." Still, I'd be spending two weeks in front of Phil trying to defend the capabilities and limitations of IBM's recently restructured software, and in front of a deservedly hostile audience of Rockwell engineers who would need to rework their expectations of it significantly.

Because of the sheer number and complexity of Shuttle hardware systems North American Rockwell had built, I was usually on the agenda at four or five meetings each day, and Phil chaired many of them. Fortunately for me, the painful lesson I'd learned from him to present my case simply and convincingly had stuck. I brought fewer charts to each session, and each one was as concise and clear as I could make it. Early the first week I was grilled with questions by Phil, and my responses were met with some skepticism by the Rockwell engineers. Still, Phil never put his cigarette down to stare at me, so I knew at least I wasn't losing ground. By the end of that week I'd gained enough credibility with him that, while he wasn't exactly defending my position, he'd begun asking more of the hard questions of the hardware engineers, especially about what changes they'd have to make on their end to get it to work. I started breathing normally again.

One of the highlights for me came near the end of the second week, though it didn't start well. An earlier meeting had run over schedule, making me late for my next one with Phil. I got to his conference room about fifteen minutes late and several agenda items into the discussion. I came in through a door at the back of the room, hoping he wouldn't notice. He did.

"You're late, Clemons," he said, startling the person who'd been presenting to him. I was about to sit down but thought better of it.

"Sorry, Phil, my last meeting ran long." I described what we'd agreed on in that one. He stared at me for a second, his cigarette hand resting on the table.

Oh, crap.

Then he told the mostly Rockwell audience, "We're starting over. There's no point in going through the agenda without getting IBM's input on the effects on the software."

He waved the engineer who'd been petitioning him for no hardware changes to sit down, and then looked at his agenda and called up the first speaker again. Phil wanted the hardware engineers to explain their stance to me? Wow! Good thing I'd taken his tutoring to heart.

The rest of that meeting and the week went well. I went home with a large number of actions required of IBM, but so did the North American Rockwell teams. I believe a good balance was struck in getting the two parties to work together, and after yet another round of laborious and tedious reprogramming by the IBM team, the software did fit in the computer's memory and ran with the speed of execution needed to complete a Shuttle mission.

As it happened, after those two weeks in California, I was seated next to Phil on the flight home. I got around to asking him if he remembered my presentation to him back in the Apollo days. He laughed. "I don't remember it was you," he said, "but I remember that session. I wondered why this guy was giving me a goddamn doctoral dissertation when all I needed was yes or no."

Error-Free Code?

By now you know John Aaron as an impressive NASA professional, but as with Phil Shaffer, his influence on me professionally was substantial. Earlier on I described how, in his role as a flight controller in Mission Control during the Apollo 12 liftoff in 1969, John saved that mission, and as a consequence, preserved the opportunity for NASA to launch every lunar mission to follow. Twelve years later John's expectations of me and the IBM team to produce the highest possible quality software for the Shuttle shaped the approach I was to take on all future projects. Though I had supported the Apollo Program, I didn't know John then. I first met him in 1981 when he was Chief of the Space Software Division at the Johnson Space Center and I was working for IBM.

As I said earlier, the Space Shuttle was a fly-by-wire spacecraft. Every action taken by a human pilot, including tugging the rotational hand controller or pressing the rudder pedals, was first converted into an electronic signal and run through the flight computer software. The computer programs evaluated, modified, and sometimes even overrode the command before sending it on to the corresponding flight equipment. In fact, the onboard computer often flew entirely without pilot involvement—for example, to stabilize the spacecraft under extremely high speed or fast maneuver conditions when human control would be impossible, as during launch.

Add to that the fact that the Space Shuttle itself was a far more complex machine than Apollo, and the challenges were compounded. The Shuttle was designed to launch into space like a winged rocket; deploy and retrieve satellites through its open bay; transport humans and supplies to and from the International Space Station, which it would also help construct; soar and lunge like a gargantuan glider through the charring heat of reentry; and be reused again and again for multiple missions spanning thirty years. On top of that, the first exhaustive test of the flight computer software was also the inaugural manned orbital flight of the Space Shuttle itself, which meant that astronaut safety was at stake during the first operational use of the code. The job of building and testing the computer programs to do all of that without fail was a daunting task.

After spending seven years as part of the IBM team of engineers and analysts responsible for the requirements of the flight programs, in June 1981 I was asked to lead the software development organization responsible for designing, implementing, and testing them. John Aaron managed the NASA organization accountable for seeing it through, and so he became my customer.

The contract that IBM had signed with NASA specified that the software we were to deliver would be "error-free"; that is to say, completely purged of all operational bugs and mistakes in coding. If you know even a little about software, you know that's virtually impossible even for moderate-sized programs. For large and complex systems it's unheard of even today. The Space Shuttle was definitely one of those. The onboard software that our team built for it comprised nearly a half-million lines of uniquely created computer instructions, each step of which was the intellectual construct of one member of the large team of IBM professionals who worked on it.

That's roughly equivalent to two million words of text, or the total combined word count of all first five volumes of George R. R. Martin's fantasy epic, *A Song of Ice and Fire.* Furthermore, while Martin labored on these volumes for fifteen years, our product, with more than two hundred authors housed in offices throughout our building, had to be ready in eight years. While revision and spelling and grammar checking are central to a work of fiction, a published book is a static object. Our final creation had to evaluate and deliberate on nonstop input from its many sensors and compute alternative strategies and reach decisions, all the while interacting with itself almost like a living entity, and then provide crucial information to people and issue rapid-fire commands to complex machinery. In other words, the flight software had to behave like a single, massively complex, intellectual construct, and not like the product of one hundred individual professionals.

I don't know how many revisions George R. R. Martin made to remove all the typos and errors from his manuscript, but it doesn't take a rocket scientist to understand that the possibility of IBM's code being error-free was unlikely. I knew I had to set up a meeting with John to talk about this "error-free code" requirement in our contract.

In July I scheduled my first official meeting as IBM's Shuttle Onboard Software System Manager with John in his role as Chief of the Space Software Division. I called ahead to tell his secretary the topic I wanted to discuss. When I showed up at his door, he smiled and welcomed me into his office. He sat down at his desk and I took a seat across from him. My first impression was a little disconcerting. John is a tall, rangy Texan. He had a quick but abbreviated smile which some callers, including me that day, mistook as amiable, but in fact it was simply cordial. He was also without artifice. His speech was measured and economical, direct and lacking pretext, and conveyed at times with a sly, understated humor in a distinctive north Texas inflection. And he was given to surprising turns of phrase.

"So the dead fish is going on the table today," he said to me, showing that slight fleeting smile. I blinked, not knowing how to respond, but from his look I quickly grasped the difference between being courteous and being chummy.

"I'm sorry, I don't understand what you mean."

"You came to tell me that IBM can't build me error-free code."

"Um . . . yes, John, I did."

I took him through the reasons why neither IBM nor anyone else could fulfill a contract like that. He listened patiently, and when I finished he looked at me for a second and then asked, "So, how many errors does IBM think you can contract for?"

"I, uh . . . I don't think I can answer that right now."

"But IBM knows that zero is the wrong answer?"

His tone was reasonable, but he wasn't smiling now. Again I wasn't sure how to respond, so I didn't say anything. He sat back in his chair.

"I understand that software has bugs, Jack, but your programs control everything on Shuttle. A code error during launch or reentry could jeopardize the mission." He let that hang in the air for a second or two and then he said, "I also know that if I require IBM to deliver code with zero errors, you'll work to get as close to that as possible."

"Error-free" stayed in our contract.

Those were more than words to John. IBM had to develop different versions of the code for each upcoming Shuttle flight, tailored to the specific needs of that mission. When a flight version was tested and verified by IBM, we'd release it for early integration with the Shuttle hardware, to North American Rockwell in Downey, California, for example, and to NASA and other contractors for use in crew training and simulations. Any code problems uncovered after IBM turned it over were counted against us, and each one lowered our contract performance measurement, and therefore our profit on the task.

In addition, I had to meet with John once a month to go over the list of those problems and answer a series of questions for each one:

What's required to fix the error?

Can the error be worked around? Making a code change to software when it's ready for flight carries its own risk.

How can IBM be sure there are no errors of a similar nature still lurking somewhere inside the code?

How did this error get through the IBM software development process, and how will IBM change its development process to prevent this type of problem on future releases?

How did this error get through the IBM independent test and verification process, and how will IBM change its testing processes to catch this type of problem on future releases?

If that sounds like an exhausting exercise, it was. Answering John's questions didn't come easy. Working to produce the level of software correctness that John Aaron expected of IBM required sweeping changes in the way we performed software development and testing on that program, and many hours of extra labor. But once we convinced John of the process modifications we made, itself a painstakingly detailed process, he reimbursed IBM for the cost to implement those he believed were justifiably in NASA's court.

He was anything but a pushover even then. One example came in the way we submitted computer programs to be run on NASA's mainframe computers. These were the days before small, high performance processors existed. As I described in the Apollo section, all code had to be typed line by line onto rectangular cards, collected into decks, and delivered by hand to NASA's Real Time Computer Complex for execution on the IBM 360 mainframe computers there. It was a tedious, time-consuming, and error-prone process, and NASA hadn't upgraded that technology since my Apollo days.

In the early 1970s IBM introduced the 3270 Display Terminal, essentially a keyboard and display system that allowed the user to communicate directly with an IBM mainframe computer. Because these displays did little else, they came to be known as "dumb terminals," but as a replacement for typing tiny holes into a set of cards fastened by rubber bands, they were a godsend. We needed to get some.

I put together a set of briefing notes, took my lead system technologist with me, and made a call on John. As usual, he sat patiently while I went through my "sales call," asked a number of technical questions, and then asked me how much it would cost to outfit our team with these terminals. I gave him a number.

He thought about it for a bit and then asked, "Will it improve your team's productivity?"

I said yes, and he asked, "How much?"

I turned to my technical lead for an answer.

"At least 10 percent," he said.

John looked at me for a minute and then said, "Okay, I'll pay for them."

I was relieved until he added, "But at the end of a year I'm going to ask you to take a cut in your manpower."

I blinked but didn't say anything, which had become a too frequent response in my meetings with him. He gave me that little smile of his.

"You just told me you're spending 10 percent of your manpower now working without these terminals, so after a year you should be able to get your job done with 10 percent fewer people. I can use that resource somewhere else."

"Um . . . okay," I said, wondering how I was going to explain to my boss how I'd agreed to a reduction in our workforce.

On our way back to the IBM building, I asked my technical lead how he came up with the 10 percent figure.

"I pulled it out of my butt," he said, without looking at me.

I was speechless. I stared at him and could see he wasn't kidding. I shook my head and groaned. "Thank god you didn't say 25 percent!"

"A Remarkable Achievement"

In the end John's expectations of IBM were warranted. His management of IBM's "error-free code" contract, and the resulting software development and test processes that IBM implemented, produced the Space Shuttle Onboard Flight Software with an extraordinarily low number of coding errors for its size. Today the industry average for supplied code is ten to twelve errors in every 1,000 lines of code. The flight software delivered for the first Shuttle launch achieved an unheard-of 0.8 errors per 1,000 lines of code, and over the thirty-year lifespan of the Space Shuttle, that error rate fell below 1.0 software error in every 5,000 lines of computer code. That's at least fifty times better than any comparable program, even today.

As good as they were, the Space Shuttle computer programs were not error-free. There were a number of critical problems found after our delivery to NASA, some of them potentially serious, and even a few that showed themselves during Shuttle flights. But at one point in 1997 there were only three errors found in the previous three versions of the software, about 1.5 million lines of code, and there were almost no errors found in the last several Shuttle flights that ended with STS-135 in 2011. A problem in the onboard code never jeopardized the spacecraft or the crew during a flight, or prevented a Shuttle mission from succeeding.

The effort wasn't cheap. In the early 1970s typical software programs cost the U.S. government about $50 per line of code. Space Shuttle

For spring '82 FSD Houston open house software verification team for onboard shuttle software showed what they do, and their creativity, in a board-game poster that follows their release-to-release validation challenge.

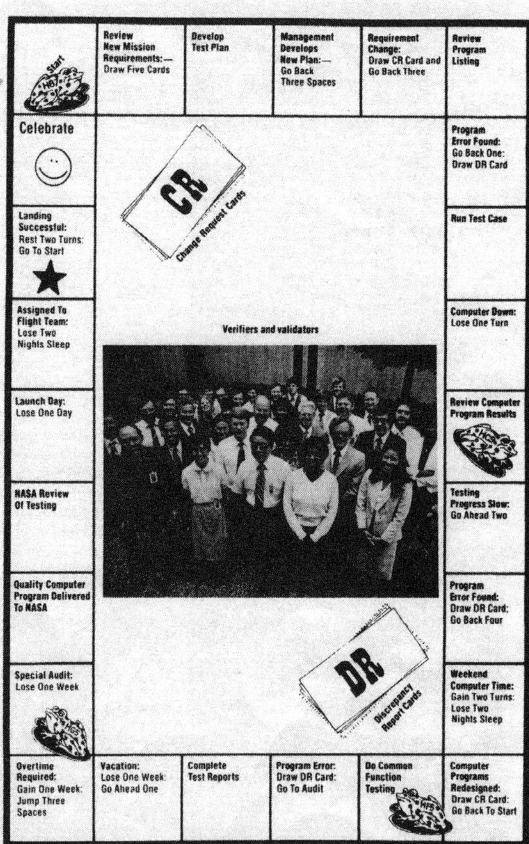

covering complex software errors. Our customer was very pleased."

Because IBM software is the pivot for everything that takes place during a Shuttle flight, the responsibility for producing "zero defects" code is a heavy one. As Lui notes: "It talks to 300 hardware boxes, responds to the crew and ground command. And it flies a rocket, a spacecraft, and an airplane combined . . . . Software has never been asked to do that before."

Add the fact that this software is built into a redundant system so that it can handle multiple system failures: "Making sure four computers perform the same tasks at exactly the same time has been a tremendous challenge to software programmers and verifiers."

But, if Oscar Lui is any example, it's a challenge they thrive on.

—AR

Oscar Lui, right, consults with onboard shuttle software verification managers Connie Parker and Dan Bowman as they prepare to test flight code.

Space Shuttle Programs FOCUS **December 1982** 5

The Shuttle Onboard Software Test team focuses on NASA's "error-free code" requirement. Image courtesy of IBM Archives.

software back then cost NASA around $1,000 per line of code. NASA paid IBM $500 million in 1976 dollars to develop, test, maintain, and support the half-million lines of code in the Space Shuttle Onboard Flight Software. That translates to about $2 billion today. Keep in mind, in an era of what are now considered primitive computing resources, most programming and testing processes were labor intensive. In the phrasing of the day: labor was cheap, hardware was expensive. If lives had been lost as a result of a flaw in the Space Shuttle software, no cost would have been too dear to have prevented it.

The Space Shuttle onboard software came closer to error-free than any large, complex software yet built. A Google search will turn up a number of technical papers and reports about how NASA and the IBM Shuttle Flight Software team and others accomplished their landmark implementation. There is justifiable pride in the people at IBM who delivered on that outstanding engineering, programming, and testing effort. In the words of one expert who assessed the Space Shuttle Program's software error frequency, Dr. Nancy G. Leveson of the Massachusetts Institute of Technology, the Shuttle onboard software was a "remarkable achievement" by those at NASA and its major contractors.

That code is still recognized as having the lowest error rate of any complex software delivered by industry before or since. But rarely do I find mentioned the person I believe was most responsible for making it happen: NASA's Chief of the Spacecraft Software Division, John Aaron. Knowing John, I'm certain he'd read this and think it's too much about him. He'd probably insist that it was a team effort by an outstanding group of people, NASA and contractors alike. That is certainly true, but John was the leader.

The Approach and Landing Tests

The Space Shuttle astronauts first tested the Orbiter's wings in the summer and fall of 1977, with a series of unpowered flights onboard the Space Shuttle *Enterprise* (yes, named to commemorate *Star Trek*'s USS *Enterprise*). These flights were designated the Approach and Landing Tests (ALT) and consisted of the *Enterprise* being carried aloft on the back of a specially outfitted Boeing 747 to an altitude of approximately

Space Shuttle Approach and Landing Tests Crews (*left to right*): C. Gordon Fullerton, Fred W. Haise Jr., Joe H. Engle, and Richard H. Truly. NASA Image.

20,000 feet and then released by means of explosive bolts to glide to a landing at Edwards Air Force Base in California. Four pilots on two alternating crews were assigned to fly a total of five so-called free flights. One of these pilots was an astronaut I'd worked with several years earlier, Dick Truly, making his first Shuttle flight.

While the Approach and Landing flights were designed to test the flight characteristics and handling qualities of the Orbiter, they were also the first operational use of IBM's onboard software. Once *Enterprise* was released from the 747 there was no going back; one way or another, it had to land. The Orbiter had no engines, no means to power the flight. It was a big, lumbering glider that had to fly-by-wire all the way to a landing. The IBM code had to work the first time and without a glitch. The flights would exercise the guidance, navigation, flight control, and displays in two ways: one with the flight software code controlling the landing, and the other with the crew manually flying the landing using computer-driven displays and cockpit window references. The Orbiter was a relatively stable airplane during its final approach to land on a runway, so this was one phase of the flight a human could fly. Even then, when a pilot moved the control stick or pedals, the flight software programs processed and modified the command before sending it along to move the Orbiter's aerodynamic flight control surfaces.

The ALT flight tests were broadcast live. My colleagues and I sat in the IBM cafeteria staring at a TV. I'm not a superstitious person, but I held my breath that day. Strangely, my biggest worry was over a mechanism I had nothing to do with. Would the explosive bolts binding the Orbiter to the 747 work properly? I had visions of the *Enterprise* getting jostled during separation and colliding with the 747's vertical stabilizer (the tail), which would definitely make for a bad day all around.

Space Shuttle *Enterprise* separates from a NASA 747 for free flight test, 1977. NASA Image.

But when separation happened it was as smooth as if moving on invisible rails. Then my anxiety shifted to the rest of the flight. That too went well. The software worked fine, including all those displays and controls I'd worked and reworked with Dick Truly's insight and guidance. Those five *Enterprise* approach and landing tests exercised the ability of the Orbiter and its onboard systems to execute an unpowered flight, maneuvering to and successfully landing on a runway. Those were five great days.

Dick Truly would later serve as pilot of the Space Shuttle *Columbia* on the second Space Shuttle launch from the Kennedy Space Center in November 1981 (designated STS-2, with STS standing for Space Transportation System). *Columbia* would also be used for the first Space Shuttle launch, STS-1, in April 1981, so Truly's second flight would mark the first time a manned spacecraft had been refurbished and returned to space.

Columbia: The First Orbital Flight

After four more years of tests and training, and constant checkout of its systems, on April 10, 1981, the Space Shuttle was finally ready to make its maiden flight into orbit. Unlike the earlier Approach and Landing flights, which involved only the Orbiter, this launch would test all of the Space Shuttle systems and therefore all of the IBM flight software programs. The stakes were high: the first-ever launch of a Space Shuttle into orbit would have two people aboard. The full capability of our software would have its first operational test on the first manned launch, a launch decision that embodied unprecedented risk. Looking back on it today, I wonder what NASA was thinking.

Space Shuttle *Columbia*, with Apollo veteran John Young and Bob Crippen (the other astronaut I'd worked with) at the controls, was ready to launch at the Kennedy Space Center. *Columbia* was mounted nose-up with an External Tank attached to its underbelly and two Solid Rocket Boosters in turn attached to the External Tank. It was Bob Crippen's first flight into space, and likewise the first mission for *Columbia* and the Space Shuttle Program. It was well before 6:00 a.m. in Houston,

Crew of Space Shuttle *Columbia* STS-1, 1981: Commander John Young (*left*) and Pilot Robert Crippen. NASA Image.

and once again my teammates and I were huddled around the television in the cafeteria.

The tension in the room was perceptible. The Orbiter's onboard computer programs were in charge of nearly everything during launch, from conducting the pre-flight checkout to igniting the main engines and SRBs, controlling the liftoff and subsequent jettisoning of the solid rockets and the External Tank once they were depleted, controlling engine burns for insertion into orbit, and continually monitoring for problems and automatically initiating one of four possible abort modes should the need arise. That was just during ascent. So we waited, knowing that the Shuttle was a complex machine and couldn't be taken for granted.

At launch minus twenty minutes, with thousands of people lining the Florida coastline and millions of viewers watching on TV, the launch was scrubbed because of a problem. Not just any problem—there was an error in our code.

John Young (*left*) and Bob Crippen training in the Space Shuttle cockpit. NASA Image.

NASA pulled the astronauts out of *Columbia* and sent the computer logs to IBM. We went to work, trying to figure out what had happened while responding to constant inquiries from NASA. The problem turned out to be a stumper. We put our most experienced software architects on it, and even with the flight data logs and everybody working around the clock, it took us more than a day to figure it out. Management's contribution was to field hourly phone calls from IBM headquarters in Armonk, New York, asking for status.

The leader of our software development team was Pat Ryk, so even though I was stressed, I knew we had the best person on the job. Pat started out as a software development programmer and worked her way forward to manage the Onboard Shuttle Software Development. Her leadership was one of the principal reasons that the flight software code came as close as it did to being error-free. I had come to depend on her perceptive and realistic assessments of the software in operation, especially when potential code problems arose during astronaut training or on launch day.

The glitch turned out to be a timing error in our software that caused it to miscommunicate with the backup computer system software developed by Rockwell International in California. That was an unacceptable occurrence, which forced Mission Control to scrub the launch. One NASA source later labeled it "the bug heard around the world," since it happened during international television coverage of the first launch of the Space Shuttle. We coded and tested a patch to the program, delivered it to NASA, and once again held our breath.

Exactly two days later, on April 12 at 7:00 a.m. EST, *Columbia* lifted off from Launch Complex 39A in Florida and began an orbital mission that marked the beginning of the Space Shuttle era. Because our software bug had canceled the first try, there was a great deal of media focus on how the onboard computer programs were behaving this time. Once again our team had the "distinction" of hearing NBC's John Chancellor and Tom Brokaw describe, live on national television, whether or not we'd done our homework. Fortunately, we had. Three days later *Columbia* executed a successful reentry and made a perfect landing at Edwards Air Force Base in California.

STS-1 was considered an overall success, though a few problems did arise. One in particular concerned the heatshield tiles that covered the undersurface of the Orbiter. During the flight the crew reported to Mission Control that two of the tiles near the nose looked as if they had "big bites out of them" and there was "significant damage" to some of the tiles near the Orbiter's tail. NASA contacted the U.S. Air Force, where classified technology was used to examine the tiles while *Columbia* was still in orbit and concluded that it wasn't a "major problem." However a later post-flight inspection of *Columbia* determined that while the Orbiter was able to make a safe reentry, sixteen tiles were lost during launch.

That U.S. Air Force examination of *Columbia* while in orbit was crucial in allowing NASA to determine the structural health of the Orbiter before its reentry—an action that might have played an even more important role had it been employed before the tragic destruction of that same Orbiter *Columbia* during reentry twenty-two years later.

Mike Mullane: No Place for Civilians

Between June 1974, when first I joined IBM in Houston, and September 1984, when I moved on from the space program, I had the chance to help design, engineer, and support seventeen flights of the Space Shuttle, including twelve launches into orbit. But one of these, the scheduled launch of Space Shuttle *Discovery* on June 25, 1984, was especially significant for me. It was to be the first spaceflight of two astronauts I'd come to know personally, one of them very well.

The area surrounding the Johnson Space Center in Texas, universally called Clear Lake City, although it wasn't officially a city and the lake was anything but clear, was a closely knit community. It occupied about twenty square miles located halfway between Houston and Galveston, and in those days the people who lived, shopped, and played there were almost exclusively employees of NASA or one of its contractors or worked in the shops and other establishments they frequented. Just as during Apollo, we shared the same neighborhoods, churches, stores, and school athletic teams. Most of us had moved there from somewhere out of state to be part of the space program. A well-worn

Astronaut Mike Mullane onboard Space Shuttle *Discovery*, STS-41D, 1984. NASA Image.

joke was the surprise we feigned at actually meeting a native-born Houstonian.

When we first arrived, Barbara and I were living in a rented duplex several miles from NASA. By the time the Space Shuttle rolled around, we were six-year residents with two young boys, and already living in our third home in a modestly upscale nearby development. I was practically an old-timer.

In January 1978 NASA selected Mike Mullane as one of a class of thirty-five astronaut candidates, a diverse group (finally) that wasn't all white males. It included six women, Sally Ride and Judy Resnik among them; three African American men, Guion Bluford, Ron McNair, and Frederick Gregory; and one Asian American man, Ellison Onizuka. All thirty-five of them later made it into space.

In 1979 Mike Mullane and I became close neighbors. When I met him, Mike was still five years away from making his first spaceflight. I don't remember how we were introduced, it might have been at a house party one of the neighbors threw, but we hit it off right away. He had a dry and somewhat indecorous sense of humor that resonated with mine. He still has it. You can get an unfiltered dose of his candid views on NASA's bureaucracy, and of the temperament of his fellow astronauts, in his 2006 memoir of his astronaut days, *Riding Rockets*.

Like me, Mike was an aerospace engineer, but he'd graduated as an Air Force officer from West Point, had flown 134 combat missions in Vietnam, and was in training as a Mission Specialist on the Space Shuttle—adventures well beyond my experiences or ambition. Since I was working for IBM, which was responsible for the software programs that ran in the Shuttle onboard computers, Mike had more than a passing interest in how that work was getting along.

Over time, our neighborliness progressed to friendship. We'd go jogging together in the evenings, at least until he grew impatient with my inability to keep up, and our families would often socialize. He had a barrel-shaped wooden hot tub in his backyard that could easily hold six adults and their wine glasses. I had a pool in my backyard, so on the sultry early evenings of Texas summer I could return the favor.

Then, as in later years, Mike was outspoken about the dangers and risks of human spaceflight. In the early 1980s NASA decided to allow a few "properly trained" civilians, people whose professional backgrounds would not otherwise qualify them as astronaut candidates, to make a flight in the Space Shuttle. In short order they announced that two members of the U.S. Congress would do so: Utah Republican Senator Jake Garn, then head of the Senate appropriations subcommittee for NASA, and Florida Democratic Congressman Bill Nelson (later a U.S. Senator), whose congressional district included the area around Cape Canaveral.

Mike was livid. He insisted that the Space Shuttle was essentially a test program, an unproven manned spaceship that was sitting atop a mass of highly volatile explosives. "Astronauts and test pilots understood the risks," he said, "but when NASA allows civilians to ride for public relations purposes, they're making an implied contract with that passenger that the shuttle system is as safe as an airplane. It is not."

In later years we all learned he was right. In April 1985 Senator Garn flew on a seven-day mission as a Payload Specialist member of the crew of the Space Shuttle *Discovery*, Mission STS-51D, the sixteenth Space Shuttle flight. Nine months later, in January 1986, Congressman Nelson flew on a six-day mission as a Payload Specialist member of the crew of the Space Shuttle *Columbia*, Mission STS-61C, the twenty-fourth Space Shuttle flight. Congressman Nelson's Orbiter touched down safely at Edwards Air Force Base on January 18. The very next flight, and just the twenty-fifth overall, was ten days later, on January 28, 1986, with the launch of the Space Shuttle *Challenger*. *Challenger* exploded seventy-three seconds after liftoff, killing all seven crew members, including civilian schoolteacher Christa McAuliffe.

That tragedy was still in the future when Mike forewarned me of the dangers. For the moment, he and I were focused on his own inaugural flight. He was assigned to make his first spaceflight as a Mission Specialist on the Space Shuttle *Discovery*, scheduled for liftoff on June 25, 1984, Mission STS-41D. As the weeks leading up to the launch date neared, Mike invited me and my family to watch his launch from the VIP viewing area at the Kennedy Space Center. He also offered to have me select something personal that he could carry into space and return as a small memento of his flight. It was a long-standing NASA tradition, and a privilege that I certainly never expected to receive. I was deeply honored and moved. I was also perplexed about what to give him, until an idea that was maybe a bit over-the-top occurred to me. Which brings me to the other crew member on his flight I knew.

Judy Resnik: Best Wishes

The other astronaut I had the chance to know personally was Judy Resnik. Not only was Judy a member of the same 1978 astronaut class as Mike—she was also scheduled to make her first spaceflight with him. Judy was a special person. She was a pilot and a classical pianist. She had a PhD in electrical engineering, and before she joined NASA she had worked as an engineer for RCA and Xerox and at the National Institutes of Health.

She was a terrific emissary for NASA. The first time we met I experienced her public speaking talent firsthand. Around 1980 or so, at Mike's urging, she came to the IBM site to talk to the engineers and programmers working on the Space Shuttle software. Astronauts sometimes made the rounds of NASA contractors to give what some of the men referred to as their "widows and orphans" sermon, as in, "if you don't do your job right, my wife will be a widow and my children will be orphans."

The speech she gave that day was electrifying. Her voice was strong and energetic; her words were assured and felt unrehearsed; her message was compelling. She described the inner workings of the code we'd just developed for the Remote Manipulator System, or RMS, the mechanical arm built in Canada that was used to catch and release satellites from inside the Space Shuttle. Judy had been in on the design of the RMS and it was evident she knew how it, and our software, should operate. She made it clear to us that she expected it to work right the first time. Later on, during her first Shuttle flight as a Mission Specialist on the Space Shuttle *Discovery* in 1984, she was one of those responsible for operating the arm, and I remembered well the admonition she gave us that day.

I had the chance to meet Judy two more times after that, both of them because of my relationship with Mike. The first was at a NASA Chili Cook-Off, held during the summer months on a grassy field inside the Johnson Space Center grounds. Invitees were restricted to NASA employees and astronauts and their families and friends.

Chili Cook-Offs, especially in Texas, were an experience unlike any other. Teams vied for prizes by preparing huge vats of steaming, and always beanless, chili with "secret" added ingredients, many invented on the spot. The judges, and all the rest of us, roved about from barrel to barrel, or cauldron to cauldron, tasting a small paper cupful of each team's brew. The sweltering air and lots of kegged beer helped keep the crowds flowing. The beer was especially useful if you happened to stumble upon one of the secret ingredients. I walked up to a barrel just as one of the "chefs" was yanking the tongue from a small brown bird (dead) and dropping it into the mix. Another half dozen of the unfortunate chickadees were strung around the barrel's rim. Or maybe it was a titmouse. I didn't linger long enough to pay my respects. I also didn't taste the sample he offered, and I swore off chili for the rest of the day.

Eventually I made my way to the astronauts' table, where Mike and the rest of his flight crew, including Judy, were cooking away. She was dressed in shorts and a white T-shirt and she looked terrific. Mike introduced me and I broke into a sweat caused by more than the Texas sun. She was very gracious. I made fumbling noises about how I appreciated her coming to speak to our team, and she smiled and pretended to remember me. By the time I'd exhausted my supply of not-so-engaging banter, one of her colleagues called to her and that was that. But I stood there thinking, beyond everything else she's done, she is really beautiful. I hoped I'd get another chance to talk to her. Mike helped make that happen too.

Barbara and I were invited to a party that Mike and his wife, Donna, were hosting at their home, and Mike mentioned that Judy would be there. This time I prepared myself; I didn't want to blather on to her like some star-struck teenager, though I have to admit, if the word "groupie" had been in fashion back then, it might have described the way I was behaving. But it was not to be. Mike's coworkers and friends were enthusiastic partiers; his house was chock-a-block full and boisterous. I glimpsed Judy once through a mass of chatting heads, but I couldn't work my way over to her side of the room, and my second chance to talk with her fluttered out of reach. I was never to meet her again, but I did have a final encounter that created a personal and lasting memory.

That brings me back to the memento Mike offered to take on his flight into orbit. As I said, it was a significant honor to me, and not as easily decided as you might expect. I didn't have an old Boy Scout badge, or some treasured childhood photograph, that felt appropriate to offer for this distinction. By then, space travel had become more personal to me, it was about the people I'd come to know and admire who were making that journey. It was that thought that finally decided me.

On Mike and Judy's first Shuttle flight together on *Discovery*, they'd each been assigned the crew position of Mission Specialist. That meant they were responsible for operating the experiments and equipment and preparing and launching the satellites to be carried into orbit on their mission. The two of them spent quite a few months of daily training together preparing for that, and so naturally, during our frequent jogs and hot tub sessions, Mike and I spent lots of time talking about her. About how bright she was. How incredibly competent and technically proficient. About her sharp, sly, and sometimes bawdy sense of humor. And, of course, how unaffectedly feminine she could be just by being herself. I didn't need to be persuaded of that.

As it happened, among his other duties, Mike was also assigned to be one of the principal photographers on his flight, which gave me the inspiration I needed. I don't remember where we were at the time, maybe jogging around our neighborhood, when I said, "I decided what I'd like you to take into space for me."

He glanced over at me, "OK, what's that?"

"A photograph of Judy," I said. Mike stopped running and so did I.

"What the hell are you talking about?"

"I mean, as my memento . . . ?"

"A photograph? There're lots of pictures of her. Don't you have an old Apollo crew patch or something you can give me?"

"I don't want just any photo, Mike. I know you'll be taking a lot of pictures while you're up there. Sometime during your flight, could you ask Judy to pose and say 'Hi, Jack' or something like that and take her picture? And after you get back, ask her to personalize it to me."

He stared at me. "That's not what NASA considers a memento."

"I understand . . . but it would mean a lot to me to have that as a remembrance of your flight."

Mike shook his head, and we started jogging again. But he didn't say no. As he said, that request was unusual. Personal tokens were typically that: an object someone already possessed that could be taken into space and returned. A personalized NASA photo of Judy Resnik in space hardly fit that description.

Mike and I went back and forth about it for several days, and in the end he said he'd talk to Judy about it, but ahead of time, well before their launch date, to see if she was okay with doing it. There was no way he was going to surprise her with my request while they were already in orbit. Graciously, Judy did agree.

The framed photo of her self-assured pose and her intelligent grin, with her hand open in a half-finished wave and a halo of weightless whisked hair surrounding her face, smiles at me now from the wall in my office. "Best wishes to Jack!" she wrote on it.

Judy's second and final flight was aboard the Space Shuttle *Challenger* on January 28, 1986.

Discovery Launch Abort: June 26, 1984

The maiden voyage of the Space Shuttle *Discovery* was originally scheduled for launch on June 25, 1984, but a software bug in the backup computer (not one of ours, I'm glad to note) delayed that launch for a day. I was there the next day, seated in the VIP viewing area at the Kennedy Space Center, with Barbara and our two sons. I'd watched two earlier manned spacecraft launches from the Kennedy Space Center: the launch of Apollo 16 to the Moon that I'd viewed while sitting on a beach blanket fourteen years earlier, and the launch of the Space Shuttle *Challenger* on April 4, 1983, when I'd taken my youngest son, Dave, to a Space Shuttle Launch to celebrate his ninth birthday. But this first-ever launch of the Space Shuttle *Discovery* in 1984 was fraught with a level of personal tension unlike either of those others, because Mike had invited us to watch the liftoff, and he and Judy were making their first spaceflight.

We sat in the bleachers with a clear view of Launch Complex 39A across five miles of flat wetlands. *Discovery* had her nose pointed at the sky, and the countdown began. Watching any launch close-up can be both exhilarating and worrisome, but especially so when you know some of the people who are on board. As I discovered that day, the stress is appreciably compounded when you also know how the systems are supposed to work and they don't do what you expect.

NASA provided guests a launch countdown narration through some nearby speakers. At T minus 31 seconds before liftoff, command was switched from Kennedy Space Center control to our onboard computers as planned. At T minus 6.6 seconds those computers commanded the Orbiter's three liquid fuel engines to ignite. These engines are started first so that they can build up thrust, which takes about six seconds, during which the Shuttle is held locked-down on the launch pad and checked for proper operation by the computers. The solid fuel Solid Rocket Boosters are ignited last since, once they are lit, there's no turning back; they will burn until their fuel is spent. At T minus 6 seconds we saw black and white smoke billow from the base of the Shuttle, meaning that the main engines had fired. Then . . . nothing.

The billowing faded, the smoke blew away, and Space Shuttle

Houston Today

IBM Federal Systems Division

Special Edition

Seventh mission starts with style after careful checkouts

5:15 a.m., June 18. . .Libby Jones, second from left, and Steve Hudson, standing, chat with Al Mandolin and "Bissy" Latoff, following night of software check-outs "just in case."

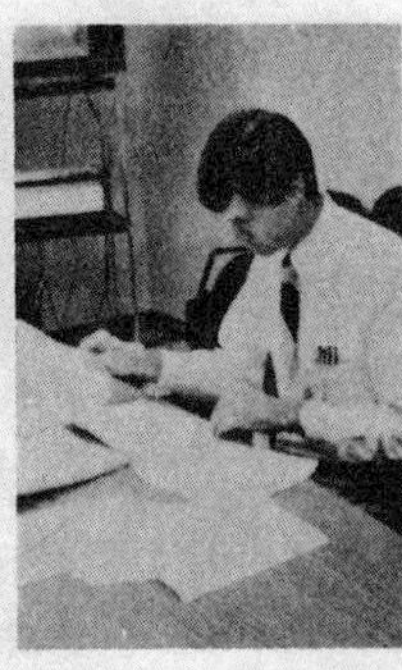

5:50 a.m. . .For Jim McGaha, a few minutes to do some post STS-7 work – scheduling field tests of software releases – before watching STS-7 liftoff.

6:15 a.m. . .Jack Clemons gets status report from Pat Ryk on audit of flight software compiler, a two-day undertaking of onboard's avionics software development and verification team.

6:30 a.m. . .In Onboard Mission Support room, Bissy Latoff and guest join Connie Parker, Pat Ryk, Jack Clemons, Tony Macina, Barbara Kolkhorst, (not shown) Ray Barton, Enrique Gomez, and Libby Jones for the liftoff of Challenger.

The Shuttle Onboard Software team preps for seventh Space Shuttle launch, June 18, 1983. Image courtesy of IBM Archives.

Discovery was still on the launch pad. Almost immediately the NASA voice monitoring the countdown announced there had been an automatic engine cutoff and the launch was aborted. I knew that meant that our software had commanded it and feared the worst, since at that point in the countdown the onboard computer was in complete control. Had a program bug caused the launch to be scrubbed, like it had on the first Shuttle flight? My stomach churned. Damn, here we go again.

We waited for a half-hour or so until NASA announced there would be no launch that day. We boarded our bus and went back to our hotel, and I spent the rest of the afternoon searching the TV for news, hoping NASA would try again the next day. But it was worse than I thought. During the aborted engine startup, a fire had broken out on the launch pad, and the crew had to be rapidly evacuated. There would be no re-launch any time soon. I called my office in Houston to find out what had happened. There was good news and bad news.

The good news was that the IBM flight software had worked correctly. It had detected a serious problem in one of the three main engines and immediately shut down the launch. That was the first time a Shuttle launch had been aborted after the main engines were ignited. It was a scary thought, and scary to watch. It was to happen four more times over the life of the program, each one due to detection of a problem and appropriate action taken by the onboard computers. But the repair for this flight turned out to be a major one. The faulty engine had to be pulled out and replaced, and the launch was going to be delayed for at least two months.

This was also the last time Barbara and I and our two sons would travel together as a family. Barbara and I had some great times together as a young couple, and later when we became the parents of two wonderful children. But in the intervening years we had grown apart and couldn't find our way back. One month later, Barbara and I filed for divorce.

First Launch of *Discovery*: August 30, 1984

Two months later, on August 30, 1984, after recovering from the dangerous engine shutdown and launch abort in June, the replacement of a faulty main engine, and an additional forty-eight-hour launch delay required to fix and test another flight software problem that had cropped up at the last minute, Space Shuttle *Discovery* and its six-astronaut crew, including Mission Specialists Mike Mullane and Judy

Resnik, were ready to try again to make their maiden voyage. STS-41D lifted off from Cape Canaveral at 8:41 a.m., and this time the launch was flawless and spectacular. *Discovery* rose into a majestic blue cloudless sky; the arc of its ascent could be seen for miles up and down the Florida coastline.

Although Mike had invited me, I wasn't there. I was back in Houston packing to move to New York. Complications in the direction my professional career was taking, and the failure of my personal life, prompted me to leave Texas and move on for good. I'd accepted a job with a different division of IBM in New York and shifted my career away from aerospace. Eight years later I returned to the complex world of large government software systems, this time air traffic control, in Maryland; but I never returned to the space program.

Launch of Space Shuttle *Discovery*, STS-41D, August 30, 1984. NASA Image.

Astronaut Judy Resnik onboard Space Shuttle *Discovery*, STS-41D, 1984. NASA Image.

Though I didn't watch Mike's liftoff from the viewing stands, his mission was an unqualified success. *Discovery*'s total cargo weighed over 41,000 pounds, the heaviest Shuttle payload up to that time. *Discovery* carried aloft three commercial communications satellites, all of which were successfully deployed. Another test item was a small prefolded experimental package that Judy Resnik opened into a large solar array. When fully extended it rose over ten stories above the Space Shuttle cargo bay. On a less auspicious note, later analysis revealed that STS-41D was the first Shuttle mission to show evidence of blow-by damage to the Solid Rocket Booster O-rings, an omen of what was to come. The virgin flight of *Discovery* returned to Earth at Edwards Air Force Base in California on September 5, 1984.

"Don't Eat Yellow Snow"

Though I'd moved on from Houston before Space Shuttle *Discovery* touched down, I had one more encounter with my old friend and neighbor, Mike Mullane. In mid-1985 the IBM group I'd joined in White Plains, New York, hosted a technology conference and invited executives from other corporations to attend. Since this wasn't long after Mike's flight on *Discovery*, I called him in Houston and persuaded him to be our featured speaker. He hadn't forgotten about my personal memento. When he came to town he brought along the photo he'd taken of Judy Resnik during their time in orbit, and that she had signed for me with her regards.

At the conference the next day we introduced Mike with some NASA footage taken during his flight, set to a recording of Elton John's "Rocket Man." Mike presented highlights and narrated photographs from his six days onboard *Discovery*, but then he turned to a story about how very happy he was that the IBM Space Shuttle computer software had worked so well on his flight. A special system of plumbing and valves had been developed for the Space Shuttle to use in orbit to dispose of liquid waste into the vacuum of space. Because of an unexpected blockage in one of the valves on *Discovery*, the effluent flowed out too slowly and accumulated into a massive frozen blob of "stuff" that had adhered to the external surface of the Orbiter.

It was a serious and even life-threatening problem. The "icicle" was a foot wide and two and a half feet long and weighed somewhere in the vicinity of thirty pounds. If it wasn't removed it could snap off during reentry and collide with the Orbiter and damage the heatshield tiles or worse. Mike told our industry executives audience that he had imagined a lot of ways he might die in space, but not one of them was getting killed by a frozen chunk of urine. The block of "ice" was finally removed by Mission Commander Hank Hartsfield, who carefully maneuvered the Shuttle's Remote Manipulator System and bumped the arm against the icicle until it broke free. The RMS itself was built in Canada, but it was controlled by an onboard computer running IBM's Shuttle flight software, and Mike was grateful that our code had worked. He reported that he was living proof that *Discovery*'s subsequent atmospheric reentry was successful. But, he added, that large block of frozen waste was likely still out there somewhere orbiting the Earth. He cautioned any alien race, traveling from some faraway star to visit and explore the Earth, to observe the old Boy Scout rule: "Don't eat yellow snow!"

6

REENTRY AND LANDING

The Amazing Flight Software People

I'd reached the final stop on my journey through spaceflight with the twelfth launch of the Space Shuttle on August 30, 1984, the maiden voyage of the Shuttle *Discovery*. I'd watched its aborted first attempt with my family at the Kennedy Space Center, and now it was the last space mission of any kind in which I had a hand. I accepted a two-year assignment with IBM in New York and moved there that September. I didn't return to Houston or work on the space program after that.

But the lessons I was taught during those sixteen years on Apollo and Space Shuttle carried me through the rest of my career. Think a problem all the way through. Focus on what's crucial and not on the extraneous. Never take a complex system for granted. Given an opportunity and a challenge, a team of experts can work miracles. And remember that getting it done doesn't matter if you don't get it right.

Now I've also reached the end of this story about the early days of the Space Shuttle. While I've tried to shed light on the inner workings of that program, like Apollo before it, at its heart this book is about the people who made it happen. Beyond the NASA leadership team and the astronauts I came to know, the hundreds of specialists at IBM who created the Shuttle Onboard Flight Software, and with whom I worked

Houston Today

IBM Federal Systems Division

Clemons receives award for workshop

Jack Clemons of FSD Houston has received an Outstanding Achievement Award for his part in the development of the division's Software Testing Workshop.

Clemons, now on assignment as a software engineering specialist in New York for the National Accounts Division, had worked three years as a Houston representative to the FSD Software Testing Coordination Group (TCOG) while the workshop was being set up.

"Everybody thought we ought to do something like this, but there hadn't been the necessary enthusiasm until Jack came along," said Rick Padinha, manager of Onboard Space Systems, at an awards ceremony held recently for Clemons.

Clemons first outlined an FSD software testing course in March 1982. "One of the objectives TCOG had two years ago was to get software testing as a discipline established within FSD, and I took on the responsibility to put together a testing course," Clemons explained.

From March 1982 through June 1984, Clemons led the development of the workshop in addition to his primary responsibility as manager of Onboard Shuttle Flight Software Development and Verification.

The workshop covers management, development testing, and software systems testing. Its objectives are: (1) to improve the quality of software testing and analysis in the division, (2) to raise awareness of current technology, which can improve efficiency and effectiveness (3) to bring testing personnel in all FSD projects a common knowledge of concepts and terminology.

During the award ceremony, Clemons said he felt the recognition should be shared with two of the workshop's pioneer instructors – Cyndy Billings, manager of Systems Software and Vehicle Cargo Systems for Onboard Space Systems, and Connie Parker, development programmer and manager of Space Station Software Engineering.

Billings and Parker helped Clemons to refine the original course outline. "Their contributions to this workshop can't be ignored," Clemons said.

(left to right) Cyndy Billings, Jack Clemons, Connie Parker

Promotion

Steven M. Eames has been promoted to manager of Communications and Community Relations, Houston, reporting to Andrew J. Cella, director of Communications.

Mr. Eames was a communications specialist, Entry Systems Division, Austin, Texas.

At Cape Canaveral:

On the move – Promotion

Brian Ehrlich from Shuttle Cargo Processing to senior associate programmer, Information Network, Tampa, Fla.

On the move – Transfer

Hank Brotherton from Business Practices and Contracts to Communications Products Division, Raleigh, N.C.

October 9, 1984 **Page 1**

Onboard Flight Software team recognized for the IBM division testing workshop (*left to right*): Cyndy Billings, me, Connie Parker. October 1984. Image courtesy of IBM Archives.

side by side every day for ten years, were the most outstanding technical people I've encountered in my career.

These were the folks who created, coded, tested, and supported the flight-critical software programs that became the Shuttle's complex digital intelligence. Some of them continued to work on the program until the last flight of *Atlantis* in 2011; others pursued different paths, and sadly, a few have died. But those with whom I stay in touch agree that the job and those people were something special. The Shuttle Onboard Flight Software folks were anything but faceless corporate drones. They were exciting, energetic, diverse, and expert. I couldn't reasonably describe all of them, but I can provide glimpses of a few more, just to give you the idea.

Earlier on, I mentioned Pat Ryk (now Pat Kolvek), who was the manager of the IBM Onboard Flight Software Development team. Pat's leadership in that role, and her focus on continually improving the IBM development process, produced the flight software quality that was outstanding. When I left the Onboard Flight Software organization in 1984, Pat replaced me as the senior manager of the entire Space Shuttle software team.

Oscar Lui is an aerospace engineer with bachelor's and master's degrees from Purdue. He and his family fled China after the Communist revolution, emigrating first to Hong Kong and later to the United States. He is both smart and extremely capable, with a natural talent for leadership. I first worked with him at TRW on Apollo, and later on the Space Shuttle at IBM. When I was appointed senior manager of the Onboard Flight Software team I named Oscar as my deputy. He had a deep understanding of the flight code and of the technical challenges of producing it. He's a person of deep-seated professional integrity, and he doggedly pursued the interests of high quality and performance with our team and with NASA's. Later he went on to manage the IBM Onboard Software Development team for the International Space Station.

Tony Macina was my colleague and mentor and is still a trusted friend. He's the finest systems engineer I've ever known. Like me, he joined IBM in June 1974 as part of the exodus of talent from TRW after it lost its Space Shuttle contract. He rose to the position of lead engineer for all of the IBM onboard Shuttle work: engineering, software development, testing, and verification. I consider him IBM's incarnation of NASA's John Aaron, both for his thoughtful and intelligent approach to a challenge and in his unruffled, low-key but always laser-focused personal style. When a young engineer or programmer came to me for career advice, I'd say, "Follow Tony on his rounds and watch how he does his job, and then try to emulate that. Or better yet, ask him to be your mentor." Later he was promoted to the level of senior manager over all of the Space Shuttle work done by IBM in Houston.

The people I've introduced here are only a few of more than one hundred professionals on this team I could sketch, people whose names are largely unknown to those not privileged to have worked beside them. But if you were to meet the people I have portrayed, and many more like them, each as uniquely qualified, you would understand why, after thirty-seven years in the company of professionals dedicated to solving difficult and thorny problems, the team on Onboard Flight Software still burns brightest to me.

The Legacy of Space Shuttle

On July 8, 2011, the Space Shuttle *Atlantis* lifted off into the grey skies over Cape Canaveral, Florida. That launch, designated STS-135, and the subsequent landing on July 21, brought to an end the ambitious and largely successful Space Shuttle Program as well as a five-decades-long era of NASA-designed manned launch vehicles stretching back to the Mercury Program and Alan Shepard's first suborbital flight in Freedom 7 on May 5, 1961. For the first time in fifty years, the United States possessed no home-grown spacecraft to transport Americans into space. Now that its era has passed, how should we measure Space Shuttle? While the Apollo Program may have been the great adventure of our age, what exactly did this stepchild do?

Simply stated, the Space Shuttle profoundly advanced the technology of spaceflight and of long-term human occupation of space. Over the Shuttle's lifetime, its crews transitioned from skilled test pilots probing the edges of a high-risk frontier to scientists and engineers building and then living and working in an orbital habitat. The Space Shuttle transformed into a demonstrated reality humanity's long and

The crew of Space Shuttle *Challenger* STS-51L died in a spacecraft explosion following launch, January 28, 1986. *From left to right, standing*: Mission Specialist Ellison Onizuka, Teacher in Space Participant Sharon "Christa" McAuliffe, Payload Specialist Gregory Jarvis, and Mission Specialist Judy Resnik; *seated*: Pilot Michael Smith, Commander Dick Scobee, and Mission Specialist Ronald McNair. NASA Image.

once seemingly unattainable dream of leaving Earth to explore new worlds. Now commercial enterprises are signing on passengers to ride into space, or to visit the Moon, or even to emigrate permanently to Mars. For those of us whose imaginations were set on fire by science fiction, we've lived to see that fabulous world become a reality.

Did Space Shuttle earn its keep? Were the advances in human spaceflight worth the costs? The Space Shuttle never achieved its operating plan of twenty-four launches a year. It significantly overran its original budget. And, of course, it suffered two horrific tragedies that might have been avoided. Yet between the first launch of *Columbia* on April 10, 1981, and the final touchdown of *Atlantis* on July 21, 2011, the Shuttle flew 135 missions over its thirty-year life. Those missions, and the astronauts who flew them, and the professionals who supported them, left us an amazing legacy. They expanded the boundaries of science and technology and the progress of human spaceflight.

One of the Space Shuttle's proud offspring is the International Space Station. The first ISS component, the Russian-built Zarya Control Module, was launched into orbit on November 20, 1998. On December 6, 1998, the in-space assembly of ISS began when Space Shuttle *Endeavour* (STS-88) delivered the U.S.-built Unity Node and then captured Zarya and mated it with Unity inside the *Endeavour*'s cargo bay. Over the next thirteen years a total of fourteen Space Shuttle launches carried components built by the United States, Canada, Europe, Japan, Brazil, and Russia for integration into the increasingly complex Space Station. ISS and Shuttle astronauts have spent more than a thousand hours in spacewalks assembling the ISS as it circles Earth in its 250-mile-high orbit. The Station is now as long and wide as a football field, and its total living and working quarters are as roomy as a five-bedroom house. It even includes an espresso machine that was custom designed for it in Italy—living in space is not quite as Spartan as it once was.

In February 2010 the Space Shuttle *Endeavour* made a night launch to the International Space Station (STS-130), the last night launch of the Shuttle Program. Though I didn't watch that one, I've had the chance to see other spacecraft launch at night. It is both a spectacular and a kind of unearthly experience. I remember reading Ray Bradbury's *R is for Rocket* as a kid and imagining what it would be like to watch a real rocket ship take off. Now I and many other people have experienced this. If you ever get the chance to see a night launch, take it.

Endeavour's mission was to deliver to the ISS a new module, named *Tranquility* after the first Apollo lunar landing site. The *Tranquility* module is fifteen feet across and the equivalent of almost two stories long, the size of a large living room. It contains an environmental support system, a gym with a treadmill and fitness training machine, an extra bathroom (can't have too many of those), and a large cupola encasing seven windows. The cupola, which was also built in Italy, gave astronauts a first-ever panoramic view of space. The view allows astronauts the visibility needed to work the remote manipulator arms outside, a necessity when attaching another new module to the Station. But the real excitement was over the unprecedented views this new "picture window" provided of the outside of the Space Station and of Earth from orbit. These sights, later transmitted to all of us, were majestic. If you've watched the movie *Apollo 13* you can see that the new views from the cupola are quite a change from the cramped old days of spaceflight. And I think we owe a debt of thanks to the Italians for contributing both luxury and grandeur to space travel.

The ISS is designed to have up to six resident crew members conducting research in the biological, physical, and space sciences and on the long-term effects of space travel on humans. Since November 2, 2000, it has never been vacant, and it has been visited by astronauts from fifteen nations. As of 2018 it has been in operation for nearly eighteen years, and more than 230 individuals have served as visiting astronauts. The focus of United States research is to determine the long-term effects of space travel on humans, a must if we are to go back to the Moon or on to other planets.

Station, as it is sometimes called, orbits about 250 miles above the Earth and makes a complete revolution every ninety minutes. NASA's website lets you find out when Station will be visible in your own sky. We've watched it pass over our house, moving from west to east, like a bright star somehow under sail. This space habitat has certainly come of age. Though the Space Shuttle's time has come to an end, the Space Station will continue operations until at least 2024.

The National Air and Space Museum in Washington regularly screens a 3D IMAX film about the Space Station, which is next best to actually being on board. If you haven't seen it, I recommend you do; it's dazzling. You can also watch videos transmitted live from the Space Station on NASA's International Space Station website. With conflicts ongoing or erupting in every part of the world, sometimes it's encouraging to see a truly international venture like Space Station working so beautifully. It has been observed that what's nice about science fiction is it assumes we will have a future. For me, the International Space Station does that too.

Another celebrated achievement of the Shuttle is the Hubble Space Telescope. It was first carried into a 354-mile-high orbit by the Space

Shuttle *Discovery* (STS-31) on April 24, 1990. It was revisited five times by the Space Shuttle between 1993 and 2009 on missions to repair, upgrade, and replace equipment and technology. The Hubble Telescope was designed for deep-space astronomical research, and people around the world have been awed by the high-quality photographic images it produced. Among these were the most striking objects ever observed, including, for example, the birth of stars thousands of light years from Earth in the Carina Nebula, which it captured in February 2010.

Beyond all the science and technology research, and placement of numerous commercial, military, and civilian payloads and satellites in orbit, these two achievements alone, ISS and Hubble, were realized only because the Space Shuttle was there to bring them into existence. These are marvels never before witnessed, and only barely imagined, in the long history of humanity. How does one place a value on the immeasurable?

So was it all worth it? Yes, I think so . . . but sadly we might have experienced all of that wonder and exploration and science without needlessly losing fourteen extraordinary people. That is the indelible stain on this otherwise impressive program.

"Space Is Hard"

On July 21, 2011, forty-two years to the day since Apollo 11 lifted off of the Moon, the Space Shuttle *Atlantis* landed on the runway at the Kennedy Space Center and brought the Space Shuttle era to an end. So where does human spaceflight go from here?

NASA has ambitious plans for its Orion Program, an off again/on again venture, a kind of Apollo Program on steroids that could carry humans back to the Moon or maybe to Mars or an asteroid. The first unmanned flight test of Orion, on December 5, 2014, was a terrific if somewhat limited success. Still, NASA has a long way to go before Orion can fulfill the potential of having humans explore an asteroid by the mid-2020s or visit Mars by the mid-2030s.

NASA's advanced Space Launch System (SLS) is the system of booster rockets that will power Orion beyond Earth orbit. This incarnation's improvement over Apollo's Saturn V is under development and, as currently designed, will not have the capability to propel humans beyond the Moon, let alone to Mars. The biggest drawback to the success of the Orion Program is a severe lack of adequate funding; it has lukewarm support from Congress, the president, and the general public. One would hope that in time common sense and sound engineering will prevail.

In September 2014 NASA awarded two contracts to private companies, one to Elon Musk's SpaceX and one to Boeing, to develop a "human-rated" spacecraft that could carry astronauts to the ISS. NASA has already named the astronaut crews who will pilot them. In spite of some early setbacks, as of this writing these companies anticipate their first flight with a human crew will launch sometime in the next one or two years. Meanwhile, the European Space Agency is busy developing its own replacement for the Space Shuttle, called the Intermediate eXperimental Vehicle (IXV).

Private corporations have stepped forward on their own to fill the gap in near-Earth orbit transport. Cargo-carrying spacecraft built by Orbital ATK, as well as by SpaceX, have made a number of successful resupply runs to the International Space Station. Richard Branson's U.S.-based company Virgin Galactic, operating out of Las Cruces, New Mexico, is working to provide suborbital flights for "space tourists" on its "air launched" winged spacecraft, SpaceShipTwo. Those are just a sampling. Jeff Bezos's company Blue Origin, for example, is planning to have his vertical takeoff and landing spacecraft, *New Shepard*, ready for human spaceflight, also in one or two years.

But success has not come easily. In October 2014 an unmanned Orbital ATK Antares mission to the ISS exploded during liftoff, and that same month Virgin Galactic's SpaceShipTwo had a disastrous accident when it broke apart in mid-air, killing one of the two pilots. In June 2015 a SpaceX Dragon spacecraft, on an unmanned ISS resupply mission, was lost when its booster rocket exploded two minutes after liftoff. Each of these companies has committed to continue with their operations, and success is already forthcoming. By the time this book is published, the list of all of the players in spaceflight, and their breakthroughs and setbacks, will be longer yet, with even more to follow.

NASA may be struggling to find its way, but private enterprise has

The crew of Space Shuttle *Columbia* STS-107 died in spacecraft disintegration during reentry, February 1, 2003. *Clockwise, from left*: Mission Specialists Kalpana Chawla and David Brown, Pilot Willie McCool, Payload Commander Michael Anderson, Payload Specialist Ilan Ramon, Mission Specialist Laurel Clark, and Commander Rick Husband. NASA Image.

found the enthusiasm and resources to take up the challenge. Perhaps the next leader to declare, "We choose to go to the Moon," and energize a generation of young scientists and engineers, will be a corporate CEO, not a United States president. The future of near-earth manned spaceflight is, as it should be, in the hands of private industry. I think we're in good hands. But it's also useful to remember that even the most convincing and credible of spaceflight plans are almost certainly optimistic. As ISS Astronaut Scott Kelly observed following the SpaceX Dragon explosion, "Space is hard."

Mission Summary

After I left Houston in 1984 and worked in New York State for eight years, I ended up in Maryland. When I landed there, it was at the invitation of my former IBM Houston Space Shuttle colleague, Rick Padinha. He'd moved to Maryland and was managing an IBM group developing new computer and software systems for air traffic control installations around the world. This I also did from 1992 to 2005, employed first by IBM and then by Loral Systems and finally by Lockheed Martin, all the while sitting in the same office in the same building in Rockville, Maryland, as the large aerospace corporations played at mergers and acquisitions. This industry I have chosen embodies its own surreal peculiarities.

At the outset I said that my youthful passion for science fiction was one of my inspirations to pursue an education that would allow me to work on the space program. Throughout my career I've worked with many professionals who, like me, have embraced a life in science or engineering because they'd seen the future as Robert Heinlein or Gene Roddenberry or William Gibson or John F. Kennedy imagined it. Now we've been to the Moon and back nine times and walked or driven on its surface six times. We've launched a fleet of winged spaceships and used them to build an orbiting multinational space station. I watched a rocket liftoff at sunset that was eerily like science fiction, backlighting the sun-stained clouds as it passed through them.

I live in a science fiction future that, amazingly, I actually helped build. Our world of technological wonders and otherworldly adventures was largely crafted by people who were inspired by dreamers. So too will the next future be.

ACKNOWLEDGMENTS

While writing is an isolating occupation, these words could not have come together without the patience, invaluable assistance, and on the mark advice of my wife Denise, herself an accomplished author.

I owe many thanks to the persistent determination of my agent, Matt Bialer, who has encouraged me for thirty years to get down on paper the stories of the people I worked with on these two great adventures. The fact that you are reading this book is only because he never gave up.

And when at last I heeded his advice, Matt found exactly the right publisher to capture it, in University Press of Florida's Deputy Director and Editor in Chief, Linda Bathgate. Linda has been enthusiastic about this book from the outset, and her advice on changes to make it more compelling are so good I wish I'd thought of them myself.

Several years ago my friend Pat Driscoll urged me to build a journal covering all the companies I worked for and the jobs I held over the course of my career, and to make notes about the leadership lessons I learned along the way. Thanks to him, that journal was invaluable in putting this book together.

Thanks to my son Paul for supplying and campaigning for the title of this book. He was so right.

Thank you to Steve Davidson, the publisher of *Amazing Stories* magazine, who gave me an early forum for blogging some of these stories, which helped me focus them and improve on how they were told.

Special thanks go my colleagues during the Apollo years, Bob Manders, Charlie Porter, and Jim Snowden at TRW Systems Group in Houston and Dave Heath at NASA, who provided their own memories for me to include, as did my IBM coworkers during the Space Shuttle era, Tony Macina, Mark Luna, and Oscar Lui.

Beyond those I talk about in this book, I would like to salute more than two hundred other professionals on the Space Shuttle Onboard Software team. Smart and extremely talented individuals who cared deeply about getting the details right, they include people like Libby Jones and Ron Kohl, Connie (Parker) Steed and B. J. Thomas, Ned Dean and Lloyd Digby, Bissy Latoff and Jim McGaha, Barbara Kolkhorst and Enrique Gomez, Cyndy (Billings) Barton and Ray Reed, Eugene McMahon, Kurt Behnke and Lin Killingbeck, Joan Shewmaker and Marie Vander Wheele, Bob Justis and Bill Ferry, Barry Eiland and Dan Bowman, Bob Harris and Lynn Small, and too many more to name. You are stars in the firmament of memory for me.

I'm equally indebted to NASA Apollo Flight Controller and Space Shuttle Orbiter Software Manager John Aaron, and to Space Shuttle Astronaut Mike Mullane, who contributed their own stories and then read the entire manuscript to make sure my memory got the facts right.

I also owe special recognition to my first wife, Barbara, who started out on this journey with me, gave birth to two remarkable sons and raised them mostly on her own, and suffered through the turbulent seas as our marriage fell apart.

My move to Maryland in 1992 to work on air traffic control systems was rewarding for me in a profound way, as it precipitated a personal transformation. My private life, the one hidden behind my "rocket science" persona, at long last completed its untidy voyage to maturity. It discovered its direction and its partner in a newly found friend who is now both my wife and my soulmate, Denise.

I've also experienced a wonderful and surprising reconnection with my two sons, Paul and Dave. Paul is a research scientist at an institute in Cambridge, Massachusetts. He and his scientist wife, Bridget, have become our close friends and occasional traveling companions. Dave is now a technology team manager working under a contract to NASA at the Johnson Space Center in Houston. He and his wife, Becky, have given us two amazing grandchildren, Nicole and Daniel, who have brightened our lives in so many ways. Those three outcomes alone make all those job and personal switchbacks worthwhile.

And of course, without those thousands of dedicated men and women whom I've never had the opportunity to meet, and who also committed their efforts and expertise to Apollo and the Space Shuttle, there would be no story to tell.

APPENDIX A

QUESTIONS PEOPLE ASK ABOUT APOLLO AND THE SPACE SHUTTLE

I lived through an era of two historic achievements and worked beside the people who gave us Apollo's conquest of the Moon and the Space Shuttle's first steps toward human beings becoming citizens of the solar system. Whether I'm chatting with new friends or giving talks to larger audiences about these programs, I find people both young and older are still fascinated by these spaceflights and hungry for more details about them. Other space program veterans have said they experience the same curiosity. So here are some of the questions I'm most frequently asked, and the sometimes little known background to the answers I provide.

What Caused the Apollo 1 Fire?

The short answer is: a combination of hazardous technology, inattentive maintenance, and flawed procedures.

By early 1966 NASA had experienced repeated successful tests of the Saturn launch vehicle, the family of booster rockets designed to take humans to the Moon. Buoyed by the accomplishment, in March 1966 NASA named the three astronaut crew members for Apollo 1, a mission into Earth orbit where they could test the equipment and procedures required by their colleagues for later flights to the Moon. The expectation was that the flight could be launched as early as February 1967. The astronauts selected for this first flight were two experienced space travelers and one rookie: U.S. Air Force Lieutenant Colonel Virgil I.

Grissom, nicknamed "Gus"; U.S. Air Force Lieutenant Colonel Edward H. White II; and U.S. Navy Lieutenant Commander Roger B. Chaffee.

Gus Grissom, a veteran pilot of both World War II and the Korean War, was one of the original seven Mercury astronauts selected by NASA in 1959. On July 21, 1961, he flew his one-man Mercury spacecraft "Liberty Bell" on a fifteen-minute suborbital flight. He was the second American to travel into space, following Alan Shepard's flight on Mercury in May of that year. Grissom's flight itself was a success, but he nearly drowned after his capsule splashed down in the Atlantic Ocean. Explosive bolts on the spacecraft's door fired unexpectedly and blew open the hatch, allowing ocean water to flood into the capsule. Grissom swam out to escape and was rescued by a recovery helicopter, but his Mercury spacecraft sank into the ocean. Following his flight, some officials accused Grissom of having accidentally fired the bolts himself, which he vigorously denied, and later investigations proved him correct.

Grissom's second spaceflight was on March 23, 1965, in the two-man Gemini 3 spacecraft. His copilot on that mission was Astronaut John Young, later to become one of the twelve men to walk on the Moon. This time Grissom's flight was a total success, and incidentally, it was the last NASA manned mission on which the entire flight was controlled by the Cape Canaveral space center in Florida. Beginning with Gemini 4 that June, operational control shifted to the recently completed Manned Spacecraft Center in Houston, Texas.

The other experienced spaceman assigned to Apollo 1 was Astronaut and Air Force Officer Ed White. White was a West Point graduate, an officer and pilot in the U.S. Air Force, and later an Air Force test pilot. In September 1962, along with Neil Armstrong, he was selected as one of the "second nine" group of NASA astronauts. White had made just one spaceflight prior to being named to the Apollo 1 crew, but it was a spectacular one. On June 3, 1965, he and his crewmate, Air Force Officer and Astronaut Jim McDivitt, began their four-day mission on Gemini 4. It turned out to be challenging for both astronauts and ground crew. One of two major objectives for the flight was for McDivitt and White to rendezvous with a depleted Titan II rocket stage that had been launched earlier that day. As Gemini 4 neared the spent rocket they discovered it was tumbling. McDivitt tried several times to approach and match orbits with the Titan II but had to give up because the maneuver was using up too much fuel.

The second objective was for Ed White to attempt the first spacewalk by a U.S. astronaut. The first-ever spacewalk, conducted by USSR Astronaut Alexei Leonov, had taken place three months earlier. About four hours into the flight both men sealed their spacesuits, Jim McDivitt depressurized the cabin, and Ed White, clutching a twenty-five-foot umbilical tether line, stepped out into the emptiness of space. For twenty minutes White floated free. Even fifty years later, the photographs and film of him—a small, human-shaped moon orbiting 150 miles above the Earth—are arresting. Concerned that White must return to the spacecraft before it entered darkness, NASA ordered him to get back in. He did, but first delayed to "take some more pictures." He reluctantly worked himself inside and closed the hatch. Ed White later confessed that having to end his spacewalk was "the saddest moment of my life."

Roger Chaffee was an Eagle Scout, a musician, an aerospace engineer, and a lieutenant commander in the U.S. Navy. He learned his love of flying from his father, who was a barnstorming pilot in the 1930s. Roger's father took him flying when the child was only seven. Roger earned his private pilot's license when he was twenty-two, and his naval aviator wings two years later. In 1962 he flew aerial reconnaissance missions from the deck of an aircraft carrier during the Cuban Missile Crisis. Chaffee was selected as one of fourteen new NASA astronauts in October 1963. Of those fourteen men, seven later went to the Moon. Four of them, including Roger, died while on duty for NASA. Had Chaffee flown on Apollo 1, he would have been the youngest U.S. astronaut to go into space.

Grissom, White, and Chaffee were not the only NASA astronauts to die during training, nor were they the first. That sad honor goes to Theodore "Ted" Freeman. He too was an aeronautical engineer, a U.S. Air Force officer and test pilot, and another member of those fourteen astronauts selected in 1963. He was killed on October 31, 1964, during a routine landing in his T38 Jet Trainer jet at Ellington Air Force Base, just north of the Manned Spacecraft Center. A goose flew into the intake of one of the aircraft engines and caused it to fail. Rather than eject, Freeman guided the plane away from some military housing and toward the air base landing strip. He died when the jet crashed into the base's perimeter fence just short of the runway.

On February 28, 1966, yet another member of that third group of fourteen astronauts, Charles ("Charlie") Bassett, and Elliot See were killed when their T38 crashed into a McDonnell Aircraft building as they attempted to land at nearby Lambert Field airport in St. Louis, Missouri. Poor visibility and bad weather in the area contributed to the crash. See and Bassett had been assigned to the Gemini 9 mission scheduled for launch three months later. They were in St. Louis to check out their spacecraft, which was being built by McDonnell Aircraft.

Just ten months after the Apollo 1 fire, on October 5, 1967, still another member of that third astronaut group, Clifton ("C.C.") Williams, also died in a T38 crash, this time in Florida and because of a mechanical failure in the jet. Williams was scheduled to be the Lunar Module Pilot on Apollo 12; had he lived he would have been the fourth man to walk on the Moon.

With these repeated accidents during training, and the loss of astronaut lives, why then did the Apollo 1 fire bring NASA to its knees? There are two reasons: most of the astronauts had flown jet aircraft for the military, and some were test pilots. The possibility of death while flying was something the members of that "right stuff fraternity" understood and accepted. T38 crashes were considered a risk of the job, and besides, that jet was just another way of getting around; it had nothing to do with getting people to the Moon.

But the fire on Apollo 1 was different. This was the very craft these space travelers would depend on to go to the Moon and return safely to Earth. It was a new vehicle, an experimental spaceship that had not yet flown. If it couldn't be trusted sitting on the launch pad during a test, then no one was safe to use it, and nobody was going to the Moon.

On January 27, 1967, NASA was running a full simulation of the Apollo 1 launch scheduled for three weeks later. The complex test required the three astronauts to be closed inside the Apollo Command Module, basically the crew's central command center, which was mounted atop the Saturn 1B rocket standing on the launch pad at the

Kennedy Space Center in Florida. The astronauts were fully suited, strapped into their seats, and the hatch of the Command Module was sealed. NASA had designated this a "plugs out" test, meaning to ensure that everything worked with the system using only its own internal power. The Mission Control team in Houston was also linked into the test using the communications systems designed for managing an actual flight.

By 6:30 p.m. the crew had been at the test for more than five and a half hours, battling through a series of nagging problems, most surrounding erratic voice communications between the Command Module and Launch Control at Kennedy. At 6:30:54 p.m. NASA noticed an electrical voltage spike in the cabin. Ten seconds later Roger Chaffee said, "Hey!" and Ed White broadcast, "I've got a fire in the cockpit!" One of the crew shouted, "There's a bad fire!" and the spacecraft's outer hull ruptured. Someone yelled, "I'm burning up!" and screamed. Then it was over. The last shout from the crew came only fifteen seconds after their first report of the fire.

Working through the smoke, heat and dust, it took workers five minutes to open the Command Module hatch from the outside. They found the bodies of the three astronauts but couldn't remove them because the fire had melted the suits into their harnesses. It was clear that two of the crew has been trying to open the hatch from the inside. Later autopsies determined that all of them had suffered third-degree burns, but by then they were already dead from asphyxiation.

On January 28, 1967, an independent investigation committee, the Apollo 204 Review Board (using NASA's internal designation of the mission), was formed to determine what had caused the tragedy. The committee comprised ten aerospace professionals drawn from inside and outside NASA, including Astronaut Frank Borman, who had flown in space for fourteen days on Gemini 7 and who would later command the Apollo 8 mission and become one of the first three humans to orbit the Moon. The committee undertook one of the most thorough and extensive examinations of an accident up to that time, including inspecting every available piece of the damaged spacecraft. When they released their final report on April 5, they concluded that the problems were both technical and, sadly, operational.

The investigation team determined that the fire itself had likely started with a short in an improperly exposed bare wire, which ignited the 100 percent pure oxygen atmosphere inside the cabin. This in turn caused the extensive Velcro covering on many of the inside surfaces, which was highly flammable in that concentration of oxygen, to combust. The fire erupted rapidly and caused a violent expansion of the oxygen, which blew out a section of the hull. The crew was unable to escape because the hatch door opened inward, making it impossible to pull it inward against the pressure inside.

If it sounds to you as though there were a lot of things wrong in that scenario, the accident committee agreed. The findings of their report eerily echo the one produced eighteen years later by the committee investigating the Space Shuttle *Challenger* accident. They state in part:

> The thorough investigation by the Apollo 204 Review Board of the Apollo accident determined that the test conditions at the time of the accident were "extremely hazardous." However, the test was not recognized as being hazardous by either NASA or the contractor prior to the accident. Consequently, adequate safety precautions were neither established nor observed for this test.
>
> The amount and location of combustibles in the command module were not closely restricted and controlled, and there was no way for the crew to egress rapidly from the command module during this type of emergency nor had procedures been established for ground support personnel outside the spacecraft to assist the crew. Proper emergency equipment was not located in the "white room" surrounding the Apollo command module nor were emergency fire and medical rescue teams in attendance.
>
> There appears to be no adequate explanation for the failure to recognize the test being conducted at the time of the accident as hazardous. The only explanation offered the committee is that NASA officials believed they had eliminated all sources of ignition and, since to have a fire requires an ignition source, combustible material, and oxygen, NASA believed that necessary and sufficient action had been taken to prevent a fire. Of course, all ignition sources had not been eliminated.

NASA fully accepted the committee's cultural assessment and all their recommended changes. For the next twenty-one months they focused exclusively on retooling and reengineering their spacecraft and transforming their work and management cultures from top to bottom. By October 1968 they were ready to try again, to launch their first manned mission, now re-named Apollo 7.

Why Were the Apollo 7 Astronauts Grounded After Their Flight?

The launch of Apollo 7 on October 11, 1968, was obviously a big one for the United States. The mission Commander was Wally Schirra, a Navy captain and one of the original seven NASA astronauts. He had flown in both the Mercury and Gemini spacecraft, and with his Apollo launch he became the first and only astronaut to fly in all three programs. Walt Cunningham, a former Marine colonel, and Donn Eisele, an Air Force colonel and test pilot, were making their first space flight, and their last.

During the course of their largely productive eleven-day mission, the crew argued repeatedly with Mission Control, refusing to obey some of their instructions, and ran afoul of NASA management as a result. In his book *Failure Is Not an Option*, Gene Kranz remembers well the Apollo 7 crew's irritable behavior on that flight. "I never figured out why Schirra had such a burr under his saddle," writes Kranz. Some of the tension may have been the stress of crewing the first manned flight of an Apollo spacecraft, or differences between the crew and Mission Control over the perceived risks of certain flight procedures, or onboard sickness.

Schirra retired immediately after his flight, and Cunningham and Eisele were never assigned to a later one. Whatever the causes, the acrimony over their behavior was such that Wally Schirra, Donn Eisele, and Walt Cunningham were the only astronaut crew members of all the Apollo flights who were not awarded Distinguished Service Medals following their mission. Schirra had received the medal twice before, one each for his Mercury and Gemini flights. NASA belatedly corrected that injustice by awarding the three men medals in 2008, though by that time only Walt Cunningham was alive to receive his.

It seems a shame, since Apollo 7 was the first manned launch of the Moon Program, and in spite of the arguing, it proved to be an important and successful test of the spacecraft and systems needed just two months later to fly humans to the Moon and back.

Was Apollo 8's Flight to the Moon Dangerous?

Apollo 8 was a bold gamble by NASA and, in hindsight, a bit reckless. The Cold War with the Soviet Union remained a major geopolitical tension. The president and the Defense Department had convinced America that there was a military advantage to landing on the Moon, and if the Russians were to get there first, the consequences could be dire. Just what that threat entailed wasn't clear. Maybe they would catapult lunar boulders at us, as in Robert Heinlein's book *The Moon Is a Harsh Mistress.*

NASA decided to pre-empt the Russians. In August 1968 the Apollo 8 Mission, originally designed to test the Apollo spacecraft in Earth orbit, was redirected to attempt the first human spaceflight to the Moon. The schedule was accelerated, the mission profile and crew training were changed, and the objective of the flight was drastically altered. The Lunar Module, the spacecraft component that would carry future astronauts to a landing on the Moon's surface, was not ready, so Apollo 8 would fly to the Moon without it. NASA's logic at the time was that the LM wasn't required, since no Moon landing was planned. The later flight of Apollo 13 in April 1970 exposed the potential hazards in that decision; an oxygen tank located in the Service Module section of that spacecraft exploded, nearly dooming the astronaut crew. They were saved only by crawling into the Lunar Module and using it as a "lifeboat" to get home. A similar explosion on Apollo 8 would have been fatal.

Apollo 8 Commander Frank Borman had been a member of the committee that investigated the Apollo 1 fire, and he led the team that re-engineered the Command Module's design afterward. He had also testified, persuasively, to a U.S. Senate committee investigating the fire, that Apollo could be made safe to fly. He understood, probably better than many, the risks of flying this unprecedented mission after only a single manned flight test of the spacecraft in orbit during Apollo 7. Like several of his colleagues, Borman was an aeronautical engineer, an Air Force fighter pilot, and a test pilot and instructor before becoming an astronaut. In 1962 he was one of the second set of astronauts selected by NASA, a group of nine that included Neil Armstrong. In December 1965 Borman and fellow Apollo 8 Astronaut Jim Lovell orbited the Earth for thirteen-plus days on Gemini 7. Borman was one of the most highly regarded astronauts.

Jim Lovell, another member of the second nine astronaut class, was a Naval Academy graduate, a fighter pilot, and a test pilot—basically the main criteria required for selection to the astronaut cadre in those days. After his flight on Gemini 7 with Frank Borman, he commanded the four-day Gemini 12 Earth-orbiting mission in November 1966. His crewmate that time was future moonwalker Buzz Aldrin. Later, in 1970, Lovell would command the ill-fated mission and rescue of Apollo 13, making him, along with John Young and Gene Cernan, one of only three humans to visit the Moon twice, and the only one of the three who didn't set foot on it. Bill Anders, yet another Naval Academy grad and fighter pilot, was making his first and only spaceflight on Apollo 8, but it was an outstanding one. In December 1968 Anders became one of the first three human beings to leave the confines of our planet. And all three returned safely to Earth.

Why Didn't Apollo 9 Go to the Moon?

Apollo 9, with Jim McDivitt, Dave Scott, and Rusty Schweickart on board, lifted off from the Kennedy Space Center on March 3, 1969. They were charged with fully assessing the new member of the Apollo family, the Lunar Module, in Earth orbit, making sure the LM was ready to fly to and land on the Moon. McDivitt and Scott were each making their second flights. McDivitt had flown on Gemini 4 and Scott on Gemini 8. Schweickart was a rookie who nevertheless was the first astronaut to earn the title Lunar Module Pilot. He and McDivitt were the two astronauts tasked to take the Lunar Module on its maiden voyage.

The ships and crew of Apollo 9 had a lot to prove during the ten days they orbited the Earth. First they had to extract the Lunar Module from the Saturn V rocket itself. During launch, the three spacecraft were stacked together on top of the rocket. Because the lunar spacecraft couldn't survive high-speed flight through the atmosphere, it was tucked with its legs folded inside a protective shroud on the uppermost section of the Saturn V. The Command and Service Module (usually referred to as CSM when they were co-joined) was mounted directly above the Lunar Module, with the CM nose pointing up and the SM engine facing down. When Apollo 9 reached orbit, the crew had to separate the CSM from the Saturn V, use their own engine to turn around and dock with the Lunar Module (which was still attached to the Saturn V upper stage), and pull it free. Mission Control then reignited the Saturn V third stage engine and sent it on a course that moved it safely away. That all worked perfectly. Objective 1: check.

Then came a series of four CSM engine firings with the LM still attached, to be sure the guidance system could handle maneuvers that increased and decreased the orbit, while everything stayed hooked together in the process. Objective 2: check.

Then the real fun began: it was time to board the LM. All three astronauts donned pressure suits and opened the hatch separating the Command Module and Lunar Module. McDivitt and Schweickart crawled into the LM while Scott remained in the CM. Schweickart powered up the LM systems and unfolded the landing gear, giving the spacecraft an insect-like appearance that earned it the crew's nickname "Spider." Beginning with the launch of Apollo 9, each crew was given permission to nickname their spacecraft to allow easy voice identification between them and Mission Control when they were flying separately. The Apollo 9 crew nicknamed the Command Module "Gumdrop" because of its shape, though "Hershey's Kiss" would have described it better.

With the two spacecraft still joined, McDivitt fired the LM's descent engine, called the Descent Propulsion System, which would later ferry astronauts to the lunar surface. At this point Schweickart felt nauseous. McDivitt advised Mission Control of his LM Pilot's illness, and they were ordered to move back into the CM and break off operations for the remainder of the day. Schweickart had experienced a form of motion sickness brought on by weightlessness. It later came to be called "space sickness" and would prove to affect more than half of all first-time space travelers. Astronauts eventually adjust to it, as did Schweickart this time, but Rusty had the misfortune of being one of the early sufferers. Frank Borman had a bout of it on Apollo 8, but it was written off as a twenty-four-hour flu. Reportedly Schweickart was quietly censured for it—these were still the days of fighter pilot "right stuff." He never received another flight assignment.

By the following day Schweickart was feeling better. He and McDivitt reentered the Lunar Module with both the CM and LM depressurized. Another task was to test the Portable Life Support System (PLSS), a specially equipped self-contained life support backpack that later astronauts would wear while on the surface of the Moon. It provided oxygen for breathing and ventilation, circulating water for body cooling, a built-in communication system, and instruments to monitor that it was working properly. Schweickart put it on, McDivitt opened the LM forward hatch, and the LM Pilot stepped out onto the Lunar Module's "front porch." At the same time Scott opened the Command Module hatch and stuck his head and upper body out to film his crewmate. The original plan called for Schweickart to make his way along the outside of both spacecraft and reenter the CM from there. The idea was to test an alternative means to transfer between the LM and CM in case the hatch joining them inside wouldn't open. That was abandoned due to concerns over Schweickart's illness the previous day. Objective 3: good job but not everything accomplished.

Schweickart gave the PLSS a thorough workout nevertheless. He spent more than forty-five minutes standing outside the LM, filming, taking photographs, gathering samples of the external surfaces for later analysis of thermal damage, and testing the LM handrail to be sure it hadn't been shaken loose by all those engine firings. All the while, he was disconnected from any source of power or oxygen or water beyond what he carried on his back. Objective 4: check. The crew called it a day. They re-pressurized both spacecraft and reassembled in the Command Module for a well-earned night's sleep; the next day they would be flying.

March 7, 1969, witnessed a milestone in aerospace history, the first manned flight of a true spacecraft, in the *Star Trek* sense of the word. Schweickart and McDivitt reentered the Lunar Module, sealed off the connecting hatch, and undocked from the CSM. The two craft flew in formation for a while, each crewman photographing the other as Spider did a slow pirouette. Then the LM used its own descent engine to move farther away. Over the next five and a half hours McDivitt and Schweickart ran through a series of maneuverings, engine firings, and orbit changes, duplicating the sequence of events needed later to descend to the Moon. At times the CSM and LM were more than one hundred miles apart, about the distance between Baltimore and Philadelphia. The LM crew were truly on their own. The final test was to ensure that the Lunar Module could return to the CSM after a Moon landing. McDivitt detonated the explosive bolts to jettison the descent engine and fired the LM ascent engine for the first time. The success of this step also ensured that Spider could work its way back to Gumdrop. Two hours later the CSM and LM were once again flying in formation and taking pictures. Then the two spacecraft successfully docked, and the first dress rehearsal for a trip to the Moon came to a close. The LM was jettisoned and left in orbit. Objective 5: two big thumbs up!

On March 13, 1969, Jim McDivitt, Dave Scott, and Rusty Schweickart's Command Module splashed down into the Atlantic Ocean in sight of the waiting recovery ship. The most complex and risk-fraught paired spaceflights yet attempted, with the Lunar Module on its inaugural flight, was a resounding success.

Did Apollo 10 Almost Crash into the Moon?

Again the short answer is: yes.

Apollo 10 lifted off from the Kennedy Space Center on May 18, 1969, heading for the Moon by way of a complex set of maneuvers and engine firings laid out as the most fuel-efficient and weight-restrictive means of getting there. The three-stage Saturn V launch rocket sequentially dropped each of its first two engine-powered sections as its fuel was spent and ignited its third powered upper stage, called the Saturn IVB (for confusion's sake), to boost the Command Module, Service Module, and Lunar Module into a circular orbit 115 miles above the Earth. The crew shut down the Saturn IVB and stayed in that orbit for about two and a half hours while final checks were made. Then the Saturn

IVB engine was restarted and fired for six minutes, pushing the spacecraft out of Earth orbit and on its way to the Moon. After the engine shut down a second time, the CSM undocked from the Saturn IVB and rotated itself to lock nose-to-nose with the LM and extract it from its shroud. Its job done, the Saturn IVB engine was fired a third and fourth time and sent on its way, slingshotting around the Moon before settling into a long, slow orbit of the Sun where it resides to this day.

With the exception of a brief midcourse correction burn of the Service Module engine, the co-joined CSM and LM coasted for three days in a long, oval-shaped orbit that would carry them to within one hundred miles of the Moon. They used the time to check and recheck all their spacecraft systems and to do a series of live TV broadcasts to Earth.

The beauty of using an orbit to go to the Moon, as opposed to just turning on the jets and burning your way straight to it, is that after you fire your engines to get you into an orbit having the shape and size you want, you don't have to fire them again to stay in it. It's basically like skiing down an icy slope and then letting your momentum carry you part way up the next one. Because there's no friction in space, when you reach the bottom of the hill in an orbit, the momentum will carry you all the way to the top again.

For safety reasons, the Apollo 10 crew used the magic of gravity to get them into a very special kind of orbit, called a free return trajectory. The shape and length of it brought Apollo 10 close enough to the Moon so that, if the crew took no action, the Moon's own gravity would grab hold of the spacecraft like a trebuchet and sling it around and back toward Earth. It was like sliding along on a track shaped like a figure-8 without needing to fire an engine, thus the name "free return." That way, if a serious problem arose in the spacecraft engines on the way out to the Moon, the interacting gravities of the Moon and the Earth would take over and send the crew home, rather than dooming them to wander forever through the solar system. It was a strategy that was crucial later, during the nearly disastrous flight of Apollo 13.

Apollo 10 had no engine problems, so when they arrived in the Moon's neighborhood, actually 95.1 miles away, the crew fired the SM engine, which was facing backward, for six minutes. That slowed them enough to fall into an orbit around the Moon instead of swinging past it. They spent the next five hours tweaking that orbit a bit and rechecking their systems and photographing the lunar surface from above. Finally they were ready to replicate the maneuvers, and hopefully the success, of the flight of Apollo 9. John Young stayed in the Command Module while Tom Stafford and Gene Cernan pressurized the Lunar Module and crawled inside. They deployed the LM landing gear, undocked from the CM, and started the series of maneuvers and descent engine burns that put them on a course to landing. Everything continued to go well, and they finally leveled out into an orbit that ranged at its lowest point to ten miles above the lunar surface. They stayed in low orbit for about three hours, taking photographs and other measurements. At one point Stafford and Cernan passed just 47,700 feet above the future landing site of Apollo 11, lower than the cruising altitude of some corporate jets.

This mission was near perfect, but that's not the same as perfect, and the difference here could have been fatal to the crew. As the Lunar Module crew prepared to separate from the descent stage and fire the ascent engine, a serious malfunction occurred. The LM began to yaw from side to side and then started rolling wildly and rapidly out of control. The erratic motion continued for nearly fifteen seconds until Stafford took control away from the onboard computer. He was able to stop the rolling motion and stabilize the spacecraft manually. Then he exploded the connecting bolts to detach the descent stage and ignited the ascent engine. The LM responded properly this time and Stafford and Cernan lifted out of their low orbit and away from the Moon and rejoined the CSM orbiting overhead.

I remember listening to the transmission from Apollo 10 when this near-calamity occurred. If that had happened during an actual liftoff from the Moon the LM would have crashed into the surface and the crew would have been lost. The initial concern was that the LM ascent guidance system had malfunctioned, not an aspect in which I was involved, but I could certainly empathize with those who were. Later analysis showed that the problem was caused by human error.

In his biography *Last Man on the Moon* Gene Cernan confessed that during the prep for ascent, he had reflexively flipped a switch from

AUTO to ATTITUDE HOLD, and a few seconds later so did Stafford, except that because it was already in that position, it got switched back to AUTO. Cernan writes: "Had we continued spinning for only two more seconds, Tom and I would have crashed."

As the terms indicate, the ATTITUDE HOLD setting would keep the LM steady, while AUTO caused the guidance to start maneuvering toward the CSM, which at that point was nowhere in sight. So it locked onto the Moon instead and tried to head there. When Stafford took over manual control he overrode the guidance and the gyrations settled out.

It's yet another example of how important it is to attend to the smallest issues, to "sweat the details," when it comes to complex systems. These machines don't think for themselves; they do exactly as they are told.

The rest of the flight of Apollo 10 proceeded well. The CSM and LM were reunited, and Tom Stafford and Gene Cernan joined John Young in the Command Module. They discarded the LM ascent stage and put it into a long orbit around the Sun. The crew fired the SM engine to leave lunar orbit and coasted "downhill" for fifty-five hours. As they neared Earth's atmosphere they discarded the Service Module to a fiery death and rode the Command Module through reentry and splashed into the Pacific a little over a mile from their planned target point. With the exception of ensuring that future crew members paid exacting attention to switch positioning, NASA was ready to place the first men on the Moon.

While Armstrong and Aldrin Were on the Moon, What Was Collins Doing?

Although Michael Collins was a critical member of the Apollo 11 crew, since he didn't walk on the Moon, he and the other lunar astronauts who served as Command Module Pilot on these flights are often overlooked. But these seven men who orbited the Moon alone, tending the store while their crewmates explored its surface, were my Apollo heroes. They were the ones who knew the CM guidance computer programs inside out, had trained for endless hours learning the countless ways the CM might behave during reentry, picked up on the subtle clues warning them that it might be going astray, and understood the often complicated actions required to take control of the flight and steer it manually to a safe and successful destination. These were the astronauts to whom our entry monitoring and backup control procedures were addressed. The information that I helped provide, and their practiced and insightful judgment, were key to returning them and their crewmates safely to the Earth.

Before his historic flight on Apollo 11, Mike Collins had been one of a two-man crew on Gemini 10. His crewmate was John Young, later to be a veteran of two flights to the Moon. Gemini 10 was the first time a human astronaut crew practiced rendezvous and docking with two separate unmanned spacecraft traveling in different orbits. Collins conducted two Extra Vehicular Activities (EVAs) on that flight, including space walking over to a spent Agena booster rocket that was drifting ten feet away. It was essential to execute these maneuvers successfully before humans could go on to the Moon.

Collins's duties as the Apollo 11 Command Module Pilot were extensive. He was in sole command of the Command and Service Module from the time the Lunar Module undocked from the CSM until Armstrong and Aldrin returned in it from the Moon. He was alone in his spacecraft for the next twenty-four hours as it orbited the Moon waiting for their return, and he was out of all contact with humankind for forty-eight minutes out of every two hours. He used the time to photograph large swaths of the lunar surface to capture necessary detail for the selection of landing sites for future missions. He was continually monitoring, and when necessary, correcting, the hundreds of mini systems that kept the CSM operating properly. "Housekeeping duties," Collins called them in his Apollo 11 memoir *Carrying the Fire: An Astronaut's Journey.*

His potentially somber duty began when the Lunar Module lifted off from the Moon. He was responsible for maneuvering the CSM and getting the two ships successfully re-docked. Or for performing a series of engine burns to rescue the Apollo 11 moonwalkers if their LM malfunctioned and they couldn't rendezvous with the CSM. Collins had worked up eighteen different scenarios for doing such a rescue, and prior to the LM lifting off to rejoin him, he restudied all of them. Finally, and most onerous of all, if the lunar astronauts were stranded on the Moon or otherwise killed, Collins would have to return to Earth alone in the Command Module. That required being prepared to operate every system on the spacecraft by himself. Fortunately neither Mike Collins nor any other Apollo Command Module Pilot was ever called upon to do so.

Flying the reentry was also the Command Module Pilot's duty, and Mike Collins handled it beautifully. In his book he admits that he had not had time in his training to practice the type of manual control takeover of an Apollo 11 reentry flight that would require a "great soaring arc after our initial penetration into the atmosphere," where the "difference between soaring an extra 215 miles and skipping out of the atmosphere altogether is slim indeed."

Just before reentry, Mission Control walked him through a refresher course on our backup procedures, but his confidence in the onboard computer and his experience simulating reentry with our monitoring procedures in hand led him to trust the automatic system to do it successfully. It was a great honor for the designers of the onboard guidance computer, folks at MIT, and for us to have earned the trust of the Command Module Pilot on Apollo 11. It was immensely gratifying to know that we didn't let him down.

Who Decided to Put American Flags on the Moon?

Today, if they think about it at all, most Americans take for granted that the Apollo 11 astronauts planted an American flag on the Moon. After all, we were the ones who put those men up there, right? What other country can say that? Well, that wasn't always the plan.

In 1967 the United States, Great Britain, and the USSR signed an Outer Space Treaty that prohibited any of these three nations from making a territorial claim to an extraterrestrial body, including the Moon (as of this writing, more than one hundred countries have ratified the treaty). As the first lunar landing of the Apollo Program neared, there was a great deal of discussion within NASA about whether planting a United States flag on the Moon would be perceived as a violation of the treaty. NASA and the president went to some lengths to make it clear that this was not our intention, but nervousness remained. Several other options were considered, including placement of a United Nations flag instead. But then, as now, the U.S. Congress was not so easily mollified.

In 1969 a House and Senate conference committee agreed on a provision in the final version of the NASA Appropriations Bill that year, stating: "the flag of the United States, and no other flag, shall be implanted or otherwise placed on the surface of the Moon, or on the surface of any planet, by members of the crew of any spacecraft . . . the funds for which are provided entirely by the Government of the United States." As a nod to the Outer Space Treaty, the bill also included the statement: "this act is intended as a symbolic gesture of national pride in achievement and is not to be construed as a declaration of national appropriation by claim of sovereignty."

As it turned out NASA had been overly concerned. There was no controversy or national protest when Neil Armstrong and Buzz Aldrin raised the American flag on the Sea of Tranquility. On the contrary, people all around the world shared in the pride of what humans had accomplished. They seemed to understand that the symbolism of raising the flag was a celebration of the people and the nation who had made it possible, much as Edmund Hillary had raised the flag of New Zealand on Mt. Everest. To my knowledge no one protested that he was proclaiming the mountain a New Zealand territory.

In addition, and with a surfeit of political caution, the Apollo 11 crew also left behind a permanent plaque displaying a map of the Earth and the inscription: "Here men from the planet Earth first set foot upon the Moon. July 1969, A.D. We came in peace for all mankind." To reinforce the sentiment, the last Lunar mission, Apollo 17, left a similar plaque with the words: "Here Man completed his first explorations of the Moon. December 1972 A.D. May the spirit of peace in which we came be reflected in the lives of all mankind."

However, the controversy isn't settled. Out of concern for protecting the historic Apollo Moon landing sites, U.S. congressional representatives have considered legislation that would designate the Apollo sites as a national park. Proponents maintain that the wording in the proposed legislation wouldn't violate the treaty, but Henry Hertzfeld of the Space Policy Institute at George Washington University disagrees. "First of all, I think it does, but secondly, even if it doesn't, other nations, including our friends and allies, are going to look at that and say, 'They're declaring sovereignty and violating the treaty.' It's going to be interpreted as yet another aggressive U.S. action." Cooler heads are hoping that the United States will work instead with the United Nations to formulate a cooperative agreement. Note also that the treaty applies to nations, not to private entities such as individuals or corporations. We'll see how this all plays out.

Planting that Apollo 11 American flag on the Moon was an engineering marvel of its own. For all its lofty symbolism, the Moon flag itself was rather unpretentious. It was an ordinary three-by-five-foot nylon American flag purchased from a U.S. government supply catalogue for $5.50. The mounting, however, was much more complicated. Pressed by the last-minute congressional edict to plant a flag, NASA turned to their own "Mr. Fix It," Jack Kinzler, the inventive engineer who would later design the protective thermal parasol used to render livable the crippled and overheated Skylab workshop. Kinzler's Lunar Flag Assembly (LFA) was designed and tested only three months before the launch of Apollo 11.

Because of the lack of atmosphere on the Moon, a horizontal telescoping crossbar was threaded through a hem sewn into the top of the flag to allow it to hang out straight, as if it were held by the wind. Due to limitations on spacecraft stowage space, and astronaut spacesuit flexibility and range of motion, both the flagpole and the crossbar were made of aluminum sections. The lower section of flagpole was supposed to be pounded into the lunar soil using a geology hammer. The upper section and flag with crossbar were then attached to that.

Every pound of cargo carried into space requires an additional pound of thrust from the rocket's engines, so mission load constraints demanded that the LFA kit itself weigh just a little over nine pounds. Another problem was the crowded conditions inside the Lunar Module, especially once the astronauts suited up to go out onto the lunar surface. Having them assemble the flag mechanisms under those circumstances proved to be unworkable. To provide the astronauts easier access, the Lunar Flag Assembly was instead folded into a protective shroud and mounted on the outside of the Lunar Module on its ladder, where it could later be retrieved by a Moon-walking astronaut.

It was tested during training and seemed to work fine, but erecting the Apollo 11 flag on the Moon was more troublesome than expected. Because the entire period of exploration on the Moon's surface for that first landing was only two and a half hours, the time involved was especially problematic. The Apollo 11 astronauts assembled the flag, but they couldn't get the horizontal section to extend all the way, which left ripples in the flag itself. They pressed the flagpole into the lunar soil, but the surface was harder than anticipated, and they could only drive it six to eight inches deep. The flag stood upright during that first lunar expedition, and photos and film show it erect and proud, and because of the ripples, it even seemed to be waving. But the pole's setting in the surface was too shallow, and the exhaust from the Lunar Module ascent engine blew the flag over during Apollo 11's liftoff from the Moon.

Are the Apollo Moon Flags Still There?

Yes, and five of those six American flags on the Moon are still standing upright. A series of photographs taken by the United States Lunar Reconnaissance Orbiter Camera (LROC) shows evidence that though the Apollo 11 flag was toppled by the exhaust from the Lunar Module ascent engine, the flags of Apollo 12, 14, 15, 16, and 17 endure. NASA launched LROC to the Moon in 2009 to make a 3-D photographic map of the Moon, and it is still in operation as of this writing. The satellite orbits the Moon continuously, taking detailed images over vast areas of the lunar surface, including repeated passes over each of the six Apollo landing sites. Scientists who have examined a series of LROC photos taken of those Apollo sites at various times of lunar day and lunar year can see the shadows of the flags circling the point where each flag was planted.

"From the LROC images it is now certain that the American flags are still standing and casting shadows at all of the sites, except Apollo 11," LROC principal investigator Mark Robinson wrote in July 2012. "I was a bit surprised that the flags survived the harsh ultraviolet light and temperatures of the lunar surface, but they did," Robinson added. "What they look like is another question."

By now their colors have been bleached away, and the fabric is probably badly tattered. Yet they stand—testaments to the greatest adventure of our age. And to the durability of items ordered from a U.S. government supply catalogue.

Could Private Citizens Track Apollo on the Way to the Moon?

Yes, every Apollo flight to the Moon and back was tracked by amateur astronomers. In fact, one private astronomy group even helped bring Apollo 13 home safely. On Thursday April 16, 1970, with the beleaguered Apollo 13 one day away from splashdown, Mission Control informed the crew they would have to do one final firing of their engines to bring the spacecraft safely back within the reentry corridor. Since this final midcourse correction would be done close to Earth, just five hours before reentry, the three NASA Deep Space Network (DSN) large dish radio antenna sites (located at Goldstone, California; Madrid, Spain; and Canberra, Australia) couldn't provide the triangulation required to get an accurate fix on its location.

Ever resourceful, NASA called on the Eastbay Astronomical Society in Oakland, California, to help. This private group of astronomy advocates, which included many seasoned astronomers, had (and still has) a long-standing relationship with Oakland's Chabot Observatory. The society had used the observatory's three large optical telescopes to track all of the Apollo flights since Apollo 7, and their reputation was such that they'd been asked to team with both private industry and NASA before this.

On Friday, April 17, Eastbay Astronomical Society got an urgent call from NASA Ames Research Station located south of them near Mountain View, California. They called their team together and went to work, using the observatory's twenty-inch refractor telescope to get a fix on the Apollo 13 CSM, and acquired the additional tracking data NASA needed. They forwarded it to NASA Ames to be relayed to the Manned Space Flight Network tracking team. Mission Control got the information they needed to perform the final engine burn accurately. It was a superb yet little known example of how even private citizens contributed to the rescue of Apollo 13.

And it's an external and verifiable nail in the coffin of conspiracy theories about NASA faking the Moon landings.

Why Is Apollo 13 Considered a Success?

In the 1995 movie *Apollo 13*, the actor Ed Harris, portraying Apollo Flight Director Gene Kranz, utters a phrase that has since entered the national zeitgeist: "Failure is not an option." It has come to represent the NASA team's "we can fix this" spirit that characterized the program in those days. The real Gene Kranz didn't say "failure is not an option" during the Apollo 13 crisis. But he appreciated the sentiment well enough to adopt it as the title of his memoir.

The events as depicted in Ron Howard's movie were the best and most visible portrayal of how everyone came together to save the crew from what had quickly escalated into a grave and deadly situation. Yet the actual NASA teamwork during those five days surpassed even that. Gene Kranz has credited the culture of "this crew is coming home!" that was demanded of his team during Apollo 13 as having arisen from NASA's failure to save the crew of Apollo 1.

Nearly every Apollo flight had a problem that could threaten the mission or jeopardize the astronauts. Each of these events taxed the team's resourcefulness and challenged them to devise a solution quickly. That was no accident. While an Apollo mission was carefully planned and repetitively rehearsed, from the outset NASA had established a team work ethic that brooked no exceptions. There would be no finger pointing; focus would be exclusively on solving the problem. Everyone involved, whether government employee or private contractor, would behave as if each wore a NASA badge. This was my first involvement in working as part of a team responsible for a critical program of this magnitude. I didn't know then that NASA's approach on Apollo was unique. Later I would discover that it was certainly not the norm for government-contractor working relationships. That experience shaped my expectations on every job to follow.

The explosion in Apollo 13's Service Module, if left unresolved, would have led to certain death for the three men aboard the spacecraft and brought about an abrupt and tragic finale to the dream of Apollo, a terrible bookend to the fire on Apollo 1. Instead, it became a triumph of human imagination and resolve, an iconic episode against which lesser efforts are still measured.

I was privileged by time and circumstances to have been a part of that effort. In the later years of my career as an engineering manager, one of the expressions I came to use with my design teams was "the system votes last." By that I mean, if the system we're designing is to work, every team member must take the same painstaking effort, expend the same tireless scrutiny, and show the same relentlessness in asking "what if?" questions. That might seem a bromide to someone who has never built a large system, but in the arenas in which I've worked, it's far more easily stated than learned. When that truism is discounted, the lessons can be severe.

Vast and incredibly intricate systems, like NASA's Apollo and Space Shuttle programs, or the FAA's Air Traffic Control modernization, comprise literally millions of complex interacting parts: machinery, computers, software, procedures, technicians, operators, and users. A complex system will never yield positive results to a mediocre approach. It's incumbent on all those building part of the system to give it their fullest attention and finest effort every day. A problem someone has overlooked while working on one small part becomes increasingly difficult to detect once it has been integrated into the whole. If something is missed, then days, months, or years down the road it will make itself known. The system will vote.

If all concerned have done their jobs, that system vote will validate us. If not, the results can be catastrophic. When a system is finally put into operation, it will act precisely in accordance with the internal logic with which it was designed and built. A complex system is impersonal; it is amoral. It is indifferent to human objectives, human aspirations, or human lives. Like reality itself, it moves only to its internal agenda.

On Apollo 13 the Service Module oxygen tank subsystem voted against the crew, but the larger system of the Apollo Program, which included years of simulation and rehearsals and training and lessons learned, had been superbly designed, and that system voted with us. That wasn't true in 1967 when three astronauts lost their lives in the fire on Apollo 1. But NASA learned from that accident and revamped and reworked Project Apollo before committing people to space. Those awful lessons and their difficult and costly corrections carried NASA through to the successful completion of Apollo 17 in 1972.

Yet in 1986 NASA encountered another system vote with the loss of Space Shuttle *Challenger* during an explosion following launch. And it happened again in 2003, with the destruction during reentry of the Space Shuttle *Columbia* and the death of its seven astronauts. In my view both these accidents could have been avoided. No one aboard *Challenger* or *Columbia* needed to die. Though potentially catastrophic system failures arose on each of these missions, the more serious failure was on the part of the Shuttle support team itself. The "failure is not an option" culture that pervaded NASA during the Apollo Program, and led to the astonishing rescue of the crew of Apollo 13, had sadly withered in the sixteen years between Apollo 13 and the loss of Space Shuttle *Challenger*.

Were the Apollo Missions Faked?

A few people today imagine that back in 1969, the U.S. government concocted an elaborate hoax meant to fool the Russians into thinking we had beaten them to the Moon; and that the whole charade was clandestinely staged in some remote corner of the Arizona desert. These silly stories started almost as soon as Apollo 11 splashed down.

Asserting that we faked it is an insult to the hundreds of thousands of men and women who worked together to accomplish the actual feat. A great body of independent data—verifiable facts obtainable from reputable sources beyond the influence of government—demonstrates the reality of the accomplishment. (The private citizens of the Eastbay Astronomical Society in Oakland, California, using the public Chabot Observatory to help in the rescue of the Apollo 13 crew, are an example of such independent sources.)

Conspiracy advocates typically claim ancient engineers using old-time technology and primitive tools couldn't possibly have pulled off various complex projects, yet the Egyptian pyramids and medieval European cathedrals still stand, and there are six American flags on the Moon.

The flag planted there by Apollo 15 astronauts Dave Scott and Jim Irwin on July 26, 1971, has earned obsessive film scrutiny from conspiracy theorists claiming that its fluttering in the "breeze" shows it is not on the airless Moon. Fact: in the vacuum of the Moon, once jostled, the flag would continue flapping for some time, exactly as seen in the film.

Astronaut Mike Mullane gave me a unique response to the conspiracy theorists. He cites another Apollo astronaut saying, "Imagine the Russians, who bankrupted their country in part by trying to win the Moon race, allowing America to fake the landing. Obviously Russia would have been the first to cry 'foul' if we had faked it. Instead, they sent their congratulations to the President."

The assorted Moon landing conspiracy myths were conclusively dismissed by the Discovery Channel *Mythbusters* team in their "Moon Landing Hoax" episode (season 6, episode 11). Yet true believers are unfazed. No dogma is more unyielding than one founded on the twin pillars of ignorance and arrogance.

What Became of the Skylab Space Station?

Skylab burned up during reentry over the Pacific Ocean in 1979. From its launch on May 14, 1973, to its descent on July 11, 1979, it undertook a magnificent, if sometimes lonely, mission as America's first human-occupied space station.

During most of 1972 TRW Systems Group in Houston had been worried that our contract to support Apollo would be terminated and we'd have to win a competitive proposal against other bidders to have a role on Skylab. That turned out not to be the case, mostly because of our extensive experience supporting the Apollo Command and Service Module astronauts during flight. For the manned portion of the Skylab Program the Command and Service Modules would be launched into orbit, perform rendezvous and docking with this new space station, and then undock and return safely to Earth, all tasks well within our sweet spot. Though TRW did suffer some loss of contract dollars for Skylab support, and therefore of money for retaining all staff, NASA had little difficulty justifying a "sole-source" contract for TRW to support this follow-on program, and no competition was required before assigning that work to us. And support it we did, from the first crew launch on May 25, 1973, until the last crew splashed down on February 8, 1974.

An idea that had been around a long time was to launch a United States manned orbiting laboratory, as it was then called, to perform science and human space endurance experiments. Werner von Braun, the German rocket scientist then working for the U.S. Army Ballistic Missile Agency, had proposed a design for one in 1959. NASA had had it on the list of future projects since the early 1960s. In 1962 a contract was awarded to the Douglas Aircraft Company in California to study the use of a converted upper stage of an Apollo Saturn V rocket, designated the Saturn S-IVB, as a manned space laboratory.

NASA had many grand plans in those days. In 1965 NASA Headquarters in Washington, D.C., established the Apollo Applications Program (AAP) to cultivate a spectrum of follow-on manned space missions using the technology developed for the Apollo Program. In addition to an Earth-orbiting laboratory, they envisioned building a permanent human-occupied outpost on the Moon and performing a manned flyby of Venus—ah, if only. Such dreams required presidential and congressional support and a lot of money, none of which was forthcoming even before Neil Armstrong walked on the Moon. NASA had requested $450 million for the Apollo Applications Program for 1967, and Congress gave them only a fraction of that; it was the smallest NASA budget since 1963. NASA canceled everything in the AAP except the Earth-orbiting laboratory, although the money to build it was not yet in the budget.

But NASA pressed on.

Skylab's own onboard computer was to be upgraded from Apollo's simple guidance system to something more state-of-the-art. The contractor selected a computer called the AP-101, built by IBM for use in U.S. Air Force jet fighters. A version of that same onboard computer would be used later for the Space Shuttle. If the Apollo Lunar Module was a lifeboat, Skylab was a luxury liner. Though it was to be built using Apollo technology, its DNA was closer to that of the later International Space Station. All in all, it was an advanced and impressive next step for human spaceflight. Then it was put on the shelf for a few years.

NASA's original plan called for a Skylab launch in 1969, but no funding from Congress was forthcoming. Then in 1970 the last four Apollo flights were canceled, which brought a silver lining to Skylab. With the Apollo spacecraft and launch vehicles already built, including the Saturn V rocket required to launch the orbital workshop itself, the budget for Skylab was suddenly affordable.

During the eight months following the rescue of the Skylab workshop in May 1973, two more teams of three astronauts lived and worked on Skylab. On July 28, 1973, a little more than a month after Pete Conrad, Joe Kerwin, and Paul Weitz splashed down, Skylab 2 was launched with astronauts Owen Garriott, Jack Lousma, and Alan Bean, another Apollo 12 veteran, on board. Their first task was to perform an EVA to install a larger twin-pole sunshade over Skylab 2's parasol to provide more extensive protection from solar heat and micrometeorites. Garriott and Lousma succeeded, with few of the EVA difficulties their colleagues had encountered on Skylab 1. They spent the remainder of their time in space continuing with the solar studies and science experiments while assessing the long-term effects of weightlessness, and of working outside in space, on human health and physiology. The crew made two more spacewalks, one by Garriott and Lousma on August 24 to replace gyroscope cables, and the second by Bean and Garriott on September 22 to make several adjustments and repairs.

The most popular members of the Skylab 2 crew weren't human. A student in a high school science class had suggested that NASA should take a spider into space to see if she could spin a web in zero gravity. The astronauts agreed and brought two spiders, named Arabella and Anita, to the workshop. Arabella had trouble spinning a web on her first try; she seemed disoriented by working in zero gravity. But she quickly adjusted, and her second attempt looked pretty much like the webs she'd spun on Earth. Later on, when Anita was given her chance, she'd already adjusted to weightlessness and her beautiful webs looked faultless from the outset. Owen Garriott transmitted a picture that he took of Arabella sitting in her perfect web, and Arabella became a TV star overnight.

After fifty-eight days of working and living in space, a period well in excess of that in any spaceflight before theirs, the Skylab 2 crew splashed down on September 25, 1973.

The third and final manned flight to Skylab, designated Skylab 3, was launched on November 16, 1973. This time three rookie astronauts, Gerald Carr, Edward Gibson, and William Pogue, made the trip. Over the course of their stay they performed studies in solar astronomy, earth sciences, physics, and human adaptation to space. They did four EVAs, totaling more than forty-one hours, to make external repairs to the spacecraft systems. They observed and recorded two first-ever events from space: the eruption of a solar flare, and the tracking of the Comet Kohoutek as it passed through the inner solar system. They also set a new space duration period of eighty-four days, a record that would stand until the International Space Station came along in 2000. When their visit was over, Carr, Gibson, and Pogue powered down the orbiting workshop, essentially putting it to sleep until another crew arrived, and returned to Earth on February 8, 1974.

Alas, no one returned to awaken it. By 1974 the Space Shuttle Program was revving up and competing for resources, so NASA decided to postpone another visit to Skylab until a Space Shuttle crew could make the journey. Unfortunately, events conspired against that: the progress on completing the Space Shuttle fell behind, while the Skylab workshop's systems began to fail and its orbit decayed more rapidly than expected. NASA looked at some options for keeping it afloat, but nothing practical could be done. On July 11, 1979, Skylab fell from the sky and broke into pieces that burned up during reentry, scattering its ashes into the Pacific Ocean near Western Australia.

What Was the Purpose of the Apollo-Soyuz Joint Mission with the Russians?

NASA had one remaining card up its sleeve for repurposing Apollo: the Apollo-Soyuz Test Project with the Russians. It would come too late for me; I'd done my last Apollo reentry support plan when Skylab 3 splashed down in 1974. The twin manned space launches by the United States and the USSR during July 1975 were less about exploration or technology or science—some of each was done—than about keeping our two nations from blowing each other up. The thirty-year Cold War between the United States and the Soviet Union had begun to thaw a little by then, though it would not melt for another fifteen years. As a measure of détente (from the French word for "relaxation"), the two superpowers agreed to perform a joint spaceflight with the U.S. Apollo Command and Service Module docking with the USSR's Soyuz spacecraft. By that time Soyuz had made eighteen manned flights into Earth orbit, although none of them had gone beyond that achievement, and four cosmonauts had died on two of those flights. Soyuz was a well-tested craft, and it is still a reliable workhorse today, carrying U.S. and Russian astronauts to and from the International Space Station.

NASA's engineers and astronauts worked closely with their Soviet counterparts to plan and rehearse the mission. NASA and the Soviets jointly designed a specialized docking module that would allow the dissimilar spacecraft to join nose-to-nose; no jury-rigged solution like the Apollo 13 CO^2 scrubber (see chapter 4) was required here. NASA had the device built in the United States.

On July 15, 1975, at 6:20 a.m. Houston time, the Soviets blasted off from their Gagarin's Start launch pad in Kazakhstan. Six and a half hours later, at 1:50 p.m., the U.S. crew lifted off at Cape Canaveral. The Russian cosmonauts were Alexey Leonov, who had made the first-ever human spacewalk on March 18, 1965, on the Voskhod 2 spacecraft, and Valeri Kubasov, a veteran of an earlier Soyuz flight.

Three American astronauts flew in the U.S. Command and Service Module. The Commander, Tom Stafford, had made three spaceflights, on Gemini 6A and 9A and on Apollo 10, in which he'd traveled to the Moon. The other two were making their first flight: Vance Brand, who would later go on to make three flights in the Space Shuttle, and a rookie who was already famous, Deke Slayton. Slayton had been one of the original Mercury Seven astronauts and the only one of them not to have flown into space. He was grounded in 1962 because of an irregular heartbeat and went on to be named Chief of the Astronaut Office, where he became one of the most important and influential figures in the Apollo Program. Slayton was responsible for selecting the crew for every Apollo Moon landing, including Neil Armstrong on Apollo 11. As an astronaut himself he was an ardent and effective advocate for ensuring the astronauts had a voice and vote on all program decision making. Read any book on the inner workings of the Apollo Program, like Gene Kranz's *Failure Is Not an Option* or James Hansen's *First Man: The Life of Neil A. Armstrong*, or watch the Ron Howard movie *Apollo 13* or the TV miniseries *From the Earth to the Moon*, and you'll come to understand how profoundly Deke Slayton shaped what became the Apollo Moon Program. In 1970 he was finally cleared for flight, and in recognition for all he had accomplished on the ground, he was assigned his first spaceflight as a member of the U.S. crew of Apollo-Soyuz. It must have been a bit awkward for Tom Stafford at first, having his long-time boss now reporting to him on this mission.

The joint launches went off without a hitch, and after a day and a half of maneuvering the two spacecraft into a common orbit, on July 17, 1975, at 11:09 a.m. Houston time the two spacecraft docked. The newly designed two-ended docking mechanism worked on the first try. Hatches on each end were opened, and the commanders Tom Stafford and Alexey Leonov worked their way into the tunnel and shook hands while millions of people around the world watched on live TV.

"Glad to see you," Stafford said to Leonov in Russian. "Glad to see you. Very, very happy to see you," Leonov answered him in English. It was an emotional moment for everyone involved.

Apollo and Soyuz remained conjoined for almost two more days, doing scientific experiments and taking photographs, several of them selfies, and then undocked. Soyuz remained in space for another five days before parachuting safely onto a remote site in Kazakhstan on July 21, on hard ground, not in the ocean, as was the custom. The Apollo Command Module spent seven more days in orbit and splashed down at 4:18 p.m. Houston time on July 24, 1975. It was the last flight of an Apollo spacecraft.

The Apollo Command Module from that flight now resides at the California Science Center in Los Angeles. The Soyuz 19 Descent Module can be seen at the RKK Energiya museum in Korolyov, Moscow

Oblast, Russia. The National Air and Space Museum in Washington, D.C., has a full-sized exhibit of the docked Apollo-Soyuz spacecraft constructed from test components and mockups of the two ships. Another one can be viewed at the Kansas Cosmosphere and Space Center in Hutchinson, Kansas. Because of their historical and international importance, they are definitely worth seeing.

How Did the Soviets Lose the Moon Race?

One of the questions I'm asked when talking about the Apollo Program is "What about the Russian astronauts who walked on the Moon?" The first time it happened I was taken aback. "The Russians didn't get to the Moon," I said. The person who asked me was visibly surprised, but how could he not know that? I was to discover he wasn't the only one. It seems some people born after the Apollo Program ended know that the United States was in a race with the Russians, and that we got there first, but do not know that the Russians didn't come in second. No other country's astronauts have even orbited the Moon. How did that happen? Weren't the Russians way ahead of us at one time? Yes, they were.

On October 4, 1957, the Soviet Union launched the world's first-ever satellite, which they had named Sputnik, Russian for "traveler." They used an intercontinental ballistic missile for the launch rocket. On September 12, 1959, the Soviets launched Luna 2, the first man-made object to reach the Moon. It was intentionally crashed onto the lunar surface. On April 12, 1961, Soviet cosmonaut Yuri Gagarin, riding in the space capsule Vostok 1 (meaning "east"), became the first human being to orbit the Earth. On March 18, 1965, Alexey Leonov in Voskhod 2 (meaning "sunrise") was the first human to leave a spacecraft and perform an EVA or spacewalk. They were well ahead of us. How could the USSR have lost the race to the Moon?

There have been volumes written on the subject, citing everything from their nation's closed culture and stifling bureaucracy to our nation's galvanizing challenge by JFK. All those aspects and more have bearing; numerous and multifaceted factors came into play. But all sources seem to agree that there were two individuals in particular whose fates provided the fulcrum for the ascendency of the U.S. program and the descent of the Soviet effort.

In the United States the turnaround began in 1957. President Dwight D. Eisenhower had grown impatient with the repeated launch failures of the U.S. Navy's Vanguard rocket. He overrode his staff's objections to having the German defector scientist, Werner von Braun, the very person who had developed rockets for the Nazis during World War II, placed in charge of ours. Eisenhower gave von Braun and his team the responsibility for the newly formed NASA Marshall Space Flight Center in Huntsville, Alabama, and he directed them to deliver a reliable rocket. In July 1960 von Braun became the center's first director and led his team to build the powerful Saturn V rocket, which went on to launch twenty-four Apollo astronauts to the Moon.

The reversal for the Soviet program happened suddenly in 1966. On January 14 Sergey Korolyov, the architect of the Soviet space program, died of a heart attack after undergoing surgery for cancer. Korolyov was the chief engineer and designer of the launch rockets and spacecraft for the Soviet Union. He led that effort from 1953 until his death. Years later his government revealed that he was the man principally responsible for the Soviet Union's success in space during those years.

In the span of a decade the influence of the two men uniquely suited for their jobs was dramatically reversed. From that point forward, regardless of later achievements and disappointments, the United States and not the USSR was destined to win the race to the Moon.

What Caused the Space Shuttle *Challenger* Disaster?

Judy Resnik was an astronaut I knew and admired. Her first spaceflight was on the Space Shuttle *Discovery* on August 30, 1984, with my neighbor Mike Mullane as one of her crewmates. Her second assignment was mission STS-51L on the Space Shuttle *Challenger*, which exploded seventy-three seconds after launch on January 28, 1986. Most people born before 1980 can remember where they were when they heard the news. I was in my office in New York when a coworker came by to tell me. I sought out a room with a television and crowded in behind thirty or forty people as the networks replayed the disaster. Even now I can't watch film of it without getting choked up. I went back to my office and called my colleagues on the onboard flight software team in Houston. They too were in shock, and nobody knew what had gone wrong.

As the day wore on we learned more about what had happened, but the exact reasons why weren't clear until the report of the Rogers Commission was released the following May (see [*Challenger* Investigation] *Report to the President by the Presidential Commission*). When I read all the "whys" I discovered that although I'd been gone from the Space Shuttle Program for little more than a year by that time, I no longer recognized the NASA described in that report.

The direct cause of the disaster was this: about one minute after launch, a seal on one of the Shuttle's Solid Rocket Boosters (SRBs) failed, spewing flames sideways and breaching the External Tank, which exploded. All seven crew members died: Judy Resnik, teacher Christa McAuliffe, Mission Commander Dick Scobee, Mission Pilot Mike Smith, Mission Specialists Ellison Onizuka and Ron McNair, and Payload Specialist Gregory Jarvis. Like Christa McAuliffe, Mike Smith and Gregory Jarvis were each making their first spaceflight. Though *Challenger* was nine miles in the air when it broke apart, later evidence indicated that at least some members of the crew may have been alive and conscious when their crew compartment impacted the ocean nearly three minutes later.

The temperature at the launch site that day ranged from 18°F overnight to 29°F at launch—below the 31°F minimum that NASA's own launch rules dictated. At temperatures below freezing, the rubber (technically fluoroelastomer) O-rings joining sections of the SRB lost some of their elasticity and might not seal properly under the stresses of launch. This effect had been understood by the engineers at Morton Thiokol, the manufacturer of the SRBs, since 1980. Similar problems had been observed in testing and reported on previous missions, including on the second Space Shuttle launch in 1981, but NASA had categorized the problem as an "acceptable flight risk," and no redesign was ordered.

So on January 28, 1986, NASA officials decided to ignore their own minimum temperature rule and proceeded with the launch of

Challenger. On the eve of the launch the Thiokol engineering team protested the decision to launch at those temperatures. They were overruled by NASA, and that decision was subsequently supported by Morton Thiokol's senior management. Why would NASA do this? The reasons aren't good.

The Rogers Commission found that the most grievous problem leading to the disaster was NASA's urgency to launch that morning. The *Challenger* launch had originally been scheduled for January 22, but a series of technical glitches and weather problems had already caused it to be rescheduled six times. *Challenger* also required a launch within a narrow time window so that two satellites in its cargo could be placed in their required orbits. That window was closing. And President Reagan's State of the Union Address was scheduled for that night, and he was expected to salute Christa McAuliffe, the first participant in his own "Teacher in Space Project."

Remember, as early as 1981 NASA had declared that this particular type of launch risk was "acceptable." By analogy, it was as if NASA was playing a logically flawed game of Russian roulette: they had pulled this trigger several times before and the gun hadn't fired, so they concluded that it must not be loaded.

The engineering team at Morton Thiokol had made a presentation to their senior management the previous day that demonstrated the high risk of O-ring failure at the predicted temperature on launch day. They strongly recommended that the launch be postponed. Unfortunately their technical arguments were somewhat confusing, and NASA was in no mood to be dissuaded. The engineers' data demonstrated that low temperatures could cause O-ring inelasticity, but their ability to articulate that cause and effect compellingly was weakened by the way it was presented. Although NASA had new information from their contractor that supported scrubbing the launch, they discounted the risk over the objections of the engineers. The engineers' charts can be reviewed in volumes IV and V of the Commission's *Challenger* investigation report.

The core of the failure was determined to be an error in human judgment and decision making, and the accident was readily preventable by simple acts of clear communication within NASA. NASA did not follow its own well-documented procedures. In making the decision to launch, NASA management had violated agency standards and launch rules. They were wrong, and the results were tragic.

I have difficulty reconciling that behavior by NASA with my own experiences of just five years earlier. I remember a series of teleconferences held between my IBM team and NASA about a problem in our flight code that was discovered days before a Shuttle launch. Our programmers had developed a fix and we were in the process of checking it out, but we hadn't yet finished all the testing. NASA repeatedly informed us that the launch would be put on hold until IBM could demonstrate to them that it was safe to proceed: exactly the opposite of what Morton Thiokol was told. What happened to change all that? I'm not sure I know.

The late Dr. Richard Feynman was a Nobel Laureate and a professor of physics at the California Institute of Technology. His reputation for clear thinking was near legendary. He had a genius intellect, was enormously curious, and possessed an exceptional analytical mind. Feynman was a member of the Presidential Commission on the Space Shuttle *Challenger* Accident. Reading the appendix he wrote to the report is a revelation. His evaluation of the gap between the Morton Thiokol engineers' analyses and NASA management decision making that contributed to the Space Shuttle *Challenger* disaster is sadly also applicable to the recurrence of tragedy on Space Shuttle *Columbia* in 2003.

"It appears that there are enormous differences of opinion as to the probability of a failure with loss of vehicle and of human life," he wrote. "The estimates range from roughly 1 in 100 to 1 in 100,000. The higher figures come from the working engineers, and the very low figures from management. What are the causes and consequences of this lack of agreement? Since 1 part in 100,000 would imply that one could put a Shuttle up each day for 300 years expecting to lose only one, we could properly ask[:] What is the cause of management's fantastic faith in the machinery? . . . For a successful technology, reality must take precedence over public relations, for nature cannot be fooled."

The loss of *Challenger* halted all Space Shuttle launches for almost three years. As after the Apollo 1 fire, NASA engaged in an intense analysis of the causes of the tragedy. The report of the Presidential

Commission on *Challenger* contained a broad set of technical and management recommendations. NASA implemented them and made a number of senior management changes. To ensure that flight safety was paramount in all Shuttle decisions, they put two of my former NASA colleagues in charge of critical operations. Astronaut Bob Crippen was named Deputy Director for Space Shuttle Operations at NASA Headquarters in Washington, D.C., and Astronaut Dick Truly was named NASA Administrator. It's probably significant that by 1995 both of them had moved on to other responsibilities. Because, unlike in the Apollo case, it seems this time those hard-learned lessons did not take permanent root.

Could the Crew of Space Shuttle *Columbia* Have Been Saved?

On February 3, 2003, near the end of the 113th flight of the Space Shuttle Program, the Shuttle Orbiter *Columbia* burned up during reentry, killing all seven astronauts onboard. The leading edge of the wings and nose of the Orbiter were covered with a reinforced carbon-carbon material that could withstand the 3000°F temperature of reentry friction. The flaw that destroyed *Columbia* had been introduced fifteen days earlier at the start of the mission, when a piece of foam insulating material on the Shuttle's External Tank broke loose during launch, impacting and seriously damaging panels of carbon heatshield material on the leading edge of the left wing, near the fuselage of the Orbiter. Launch video taken at the time showed the insulation break away and strike the Orbiter, but NASA chose to take no action. *Columbia*'s wing burned through and melted off during reentry.

External Tank insulation and other material had broken away on four previous Shuttle flights beginning as early as 1988, but since damage to the Orbiter heatshield was "minor" in those cases, NASA concluded that the Shuttle was safe, and this too became acknowledged as an "acceptable risk." As noted, the aerodynamic heat that builds up around Shuttle during reentry can reach 3000°F, not as high as Apollo's 5000°F, but still hot enough to melt iron. The damaged inner section of *Columbia*'s wing was unable to deflect that heat, which burned through to the fuselage and destroyed the spacecraft.

Another committee was formed, the *Columbia* Accident Investigation Committee, to look into the causes of the tragedy, and just as after the *Challenger* accident, the findings were depressing. Once again the commission concluded that NASA management's judgment and decision-making culture were seriously flawed. Although evidence of tile and wing damage had been identified on earlier flights, no design change had been made, based on the logic that "the damage wasn't serious before, so it must not be a big risk."

The unsound logic of NASA's Russian roulette led to the death of seven more astronauts: Mission Commander Rick Husband, Mission Pilot William McCool, Payload Commander Michael Anderson, Mission Specialists David Brown, Kalpana Chawla, Laurel Blair Salton Clark, and Payload Specialist Ilan Ramon, who was the first astronaut from the nation of Israel. William McCool, David Brown, Laurel Clark, and Ilan Ramon were each making their first spaceflight. Reading this accident report is profoundly disheartening (see *Columbia Accident Investigation Board*, chapter 6).

In his 2006 memoir *Riding Rockets*, Astronaut Mike Mullane reported that on his own STS-27 classified mission on Space Shuttle *Atlantis* in 1988—the second flight after the *Challenger* disaster in 1986—the crew reported to Mission Control that something had broken away during launch and impacted the Orbiter. Each of the Shuttle's Solid Rocket Boosters had an Apollo heatshield-like ablative material on its nose to provide protection from the heat load during ascent as the vehicle went supersonic in the lower atmosphere. On STS-27 some of that ablative material on the right side SRB broke off and damaged the tiles. During that flight NASA responded to the crew that the damage "was not a significant concern." However a post-flight inspection of *Atlantis* showed that almost seven hundred tiles had been damaged, extending along the length of the Orbiter's underbelly, and there was a partial melt-through of the now unprotected metal structure underneath. Only a narrow breadth of partially melted aluminum separated that flight from disaster. Sound familiar?

Not everyone working on the ill-fated *Columbia* flight had been sanguine. Some of the NASA engineers who viewed the launch video showing material striking the Orbiter were concerned that the damage was serious and alerted their management. But NASA management concluded that even if this was a problem, nothing could be done, so no further action would be taken. The engineers pressed to have the Department of Defense use their classified technology to examine *Columbia*'s wings and underbelly for damage. A similar request of the Air Force had been made by NASA during the first Space Shuttle flight in 1981, and the Air Force had complied and examined the damage, which was found to be minor. NASA management refused to make the request this time, again citing their rationale that nothing could be done to save the crew. So NASA proceeded with the mission as if the danger was manageable. As with *Challenger*, they were wrong. Seven of the best and brightest our world had to offer paid with their lives.

The *Columbia* Accident Investigation Committee also challenged NASA's contention that nothing could have been done. They evaluated a number of potential rescue alternatives and found two that might have been workable: a rescue mission by the Space Shuttle *Atlantis*, which was then inside the Kennedy Space Center Vehicle Assembly Building being readied for launch in early March; and an emergency spacewalk to attempt to repair the damaged edge of the left wing. Though both approaches were risky and success wasn't assured, they might have been attempted if NASA had acted soon enough.

But NASA didn't act. The "failure is not an option" culture that we saw during the Apollo Program, and that led to the astonishing rescue of the crew of Apollo 13 and the repair of Skylab, had sadly withered in the thirty years between Apollo 13 and the loss of Space Shuttle *Columbia*. Astronaut "Hoot" Gibson, a veteran of five Shuttle missions, including as commander of that near-disastrous *Atlantis* STS-25 mission in 1986 with Mike Mullane aboard, offered his observations in a 2009 interview (see *Legendary Commander Tells Story of Shuttle's Close Call*):

"NASA does amazing things when they've got their back against the wall," he said. "Like Apollo 13. I've seen us work out some really dramatic things in some of the missions when we had on-orbit problems and we did in-flight maintenance and things like that. You never know what you could have done because you didn't try."

APPENDIX B

FEATURED MISSIONS

Apollo

Apollo 1: January 27, 1967

Objective: Planned first manned launch into Earth orbit, did not launch, crew perished in fire in Command Module during a test on the launch pad

Crew:

Commander Virgil I. "Gus" Grissom
Command Module Pilot Edward H. White II
Lunar Module Pilot Roger B. Chaffee

Apollo 8: December 21–27, 1968

Objective: First manned spacecraft to orbit the Moon

Crew:

Commander Frank F. Borman, II
Command Module Pilot James A. Lovell Jr.
Lunar Module Pilot William A. Anders

Apollo 11: July 16–24, 1969

Objective: First spacecraft to land humans on the Moon

Crew:

Commander Neil A. Armstrong

Command Module Pilot Michael Collins
Lunar Module Pilot Edwin E. "Buzz" Aldrin Jr.

Apollo 13: April 11–17, 1970

Objective: Planned third manned landing on the Moon, explore the Fra Mauro region. Mission was aborted following an explosion in the Service Module during the flight out to the Moon; crew was returned to Earth safely

Crew:

Commander James A. Lovell Jr.
Command Module Pilot John L. Swigert
Lunar Module Pilot Fred W. Haise Jr.

Skylab 1: May 25–June 22, 1973

Objective: First manned crew to occupy the Skylab space station

Crew:

Commander Charles "Pete" Conrad Jr.
Science Pilot Joseph P. Kerwin
Pilot Paul J. Weitz

Space Shuttle

Approach and Landing Test Free-Flights

Objective: Test the flight behavior of the Orbiter during a gliding approach and landing

Free-Flight 1: August 12, 1977

Crew:

Commander Fred W. Haise Jr.
Pilot C. Gordon Fullerton

Free-Flight 2: September 13, 1977

Crew:
Commander Joe H. Engle
Pilot Richard H. Truly

STS-1 Space Shuttle *Columbia*: April 12–14, 1981

Objective: First launch, orbit, and landing of the Space Shuttle Program
Crew:
Commander John W. Young
Pilot Robert L. Crippen

STS-41D Space Shuttle *Discovery*: August 30 1984–September 5, 1984

Objective: 12th flight of Space Shuttle Program, first flight of Space Shuttle *Discovery*
Flight Crew:
Commander Henry W. Hartsfield Jr.
Pilot Michael L. Coats
Mission Specialists:
Richard M. Mullane
Steven A. Hawley
Judith A. Resnik
Payload Specialist:
Charles D. Walker

STS-51L Space Shuttle *Challenger*: January 28, 1986, exploded 73 seconds after liftoff; all aboard perished

Objective: 25th flight of Space Shuttle Program, deploy Tracking and Data Relay Satellites, scientific observation of Halley's Comet, transport a high school teacher into space as part of Teacher in Space Project

Flight Crew:
- Commander Francis R. Scobee
- Pilot Michael J. Smith

Mission Specialists:
- Ellison S. Onizuka
- Judith A. Resnik
- Ronald E. McNair

Payload Specialists:
- Gregory B. Jarvis
- S. Christa McAuliffe (teacher)

STS-107 Space Shuttle *Columbia*: January 16–February 1, 2003, destroyed during reentry, all aboard perished

Objective: 113th flight of Space Shuttle Program, first flight and use of the pressurized SPACEHAB Research Module mounted in the Shuttle Payload Bay, performed numerous scientific experiments for both U.S. and international partners with much of the data transmitted to Earth during the mission

Flight Crew:
- Commander Rick D. Husband
- Pilot William C. McCool

Mission Specialists:
- David M. Brown
- Kalpana Chawla (India)
- Laurel B. Clark

Payload Commander:
- Michael P. Anderson

Payload Specialist:
- Ilan Ramon (Israel)

GLOSSARY OF TERMS AND ACRONYMS

ALT—Space Shuttle Approach and Landing Test flights

CapCom—Capsule Communicator, usually an astronaut; the only person in Mission Control authorized to communicate directly with the crew of an Apollo or Space Shuttle mission in progress

CDR—Spacecraft Commander Apollo/Space Shuttle

CM—Apollo Command Module

CMP—Apollo Command Module Pilot

CSM—Apollo Command Module and Service Modules when co-joined

DSN—Deep Space Network, three large dish radio antenna sites located at Goldstone, California; Madrid, Spain; and Canberra, Australia; used by NASA to track distant spacecraft

DSKY—Apollo Command Module onboard Display and Keyboard device

EECOM—Electrical, Environmental and Communications Systems

Entry Corridor/Reentry Corridor—The narrow passageway through the atmosphere which a spacecraft must travel during reentry to avoid either failing to reenter at all on the upper side or experiencing excessive g-forces on the lower edge

EI—Entry Interface, start of an Apollo reentry, defined as 400,000 feet above Earth's surface

ET—External Tank, large external liquid fuel tank mounted underneath the Orbiter used to power the Space Shuttle Main Engine during launch; discarded after launch

EVA—Extra Vehicular Activity, a spacewalk or a walk on the Moon

g-force (g's)—Additional force experienced by an astronaut's body during launch and reentry measured in multiples of his or her body weight

JSC—NASA Johnson Space Center in Houston, Texas (Manned Spacecraft Center prior to 1973)

KSC—NASA John F. Kennedy Space Center at Cape Canaveral, Florida

LM—Apollo Lunar Module

LMP—Apollo Lunar Module Pilot

Mach number—Aircraft or spacecraft speed measured in multiples of the speed of sound moving through Earth's atmosphere (761 miles per hour at sea level)

MSC—NASA Manned Spacecraft Center in Houston, Texas (renamed Johnson Space Center in 1973)

MSFN—Apollo Manned Space Flight Network, Earth-based

NASA—National Aeronautics and Space Administration (established in 1958)

NACA—National Advisory Committee for Aeronautics (dissolved in 1958)

Orbiter—The aircraft-like component of the Space Shuttle

RTCC—Real Time Computer Complex at NASA Johnson Space Center during Apollo and Space Shuttle Programs

SATURN V—NASA liquid fuel rocket with sufficient power to launch Apollo astronauts to the Moon

"SCE to AUX"—Put the Apollo Command Module Signal Conditioning Electronics switch into its Auxiliary position.

SM—Apollo Service Module

SRB—Solid Rocket Booster, one of two external solid fuel rockets used to launch the Space Shuttle; discarded after launch

STS—Space Transportation System, the official name of the Space Shuttle Program, also used as a Shuttle Mission designator, STS-1 for example

Task Monitor—NASA employee responsible for a private contractor's work progress on a task

Task Manager—Contractor employee leader responsible for work progress on a task

X-15—Early experimental rocket-plane jointly developed by NASA and U.S. Air Force

REFERENCES AND FURTHER READING

The Apollo Flight Journal (all flights, Apollo 7 to Apollo 17). David Woods, NASA History Division, n.d. https://history.nasa.gov/afj/.

The Apollo Lunar Surface Journal (Apollo 11, 12, 14, 15, 16, and 17). Eric M. Jones and Ken Glover, eds., NASA History Division, n.d. https://www.hq.nasa.gov/alsj/frame.html.

Apollo 10, The Fourth Mission: Testing the LM in Lunar Orbit 18 May–26 May 1969. NASA History Office, n.d. https://history.nasa.gov/SP-4029/Apollo_10a_Summary.htm.

Apollo 11, Moon Landing Footage Out-Take. Snopes.com http://www.snopes.com/inboxer/hoaxes/moonhoax.asp.

The Apollo 13 Flight Journal. David Woods, Alexandr Turhanov, and Lennox J. Waugh, NASA History Division, n.d. https://history.nasa.gov/afj/ap13fj/index.html.

Apollo 13 Lunar Surface Journal. Eric M. Jones, NASA History Division, n.d. https://www.hq.nasa.gov/office/pao/History/alsj/a13/a13.html.

Apollo 15 Flag Deployment (discussion of origin and unique size of Apollo 17 Flag). Eric M. Jones and Ken Glover, eds., NASA History Division, 2012. https://www.hq.nasa.gov/alsj/a15/a15FlagDeployment.html.

The Apollo 17 Lunar Surface Journal: ALSEP Off-load. Eric M. Jones and Ken Glover, eds., NASA History Division, n.d. https://www.hq.nasa.gov/office/pao/History/alsj/a17/a17.alsepoff.html.

Apollo Lunar Surface Journal: Apollo 15 Flag Deployment. James Fincannon, April 21, 2012. https://www.hq.nasa.gov/alsj/a15/a15FlagDeployment.html.

APOLLO Mission 1. Steve Garber, NASA History Web Curator, NASA History Office, September 10, 2015. https://history.nasa.gov/Apollo204/.

Belew, Leland F., ed. *Skylab, Our First Space Station*, ch. 4: "Rendezvous and Repair." George C. Marshall Space Flight Center, NASA Scientific and Technical Information Office, 1977. https://history.nasa.gov/SP-400/ch4.htm.

Benson, Charles D., and William Barnaby Faherty. *Moonport: A History of Apollo Launch Facilities and Operations: Pruning the Apollo Program.* NASA Special Publication-4204, NASA History Series, 1978. https://www.hq.nasa.gov/pao/History/SP-4204/ch22-8.html.

Cernan, Eugene, and Don Davis. *Last Man on the Moon.* New York: St. Martin's Press, 1999.

[*Challenger* Investigation]. *Report to the President by the Presidential Commission on the Space Shuttle Challenger Accident,* vol. 1. Washington, D.C.: NASA, June 6, 1986. https://spaceflight.nasa.gov/outreach/SignificantIncidents/assets/rogers_commission_report.pdf.

Clemons, J. F., and P. E. Moseley. Recommended Entry Monitoring and Backup Control Procedures for Apollo 11 (Mission G). NASA MSC Internal Note No. 69-FM-185, July 2, 1969.

Collins, Michael. *Carrying the Fire: An Astronaut's Journey.* New York: Farrar, Straus and Giroux, 2009.

Columbia Accident Investigation Board. *Report of Columbia Accident Investigation Board,* vol. 1, ch. 6: "Decision Making at NASA." Washington, D.C.: NASA, August 26, 2003. http://s3.amazonaws.com/akamai.netstorage/anon.nasa-global/CAIB/CAIB_lowres_chapter6.pdf.

Computers in Spaceflight: The NASA Experience, ch. 2: "The Apollo Guidance Computer: Hardware." NASA History Office, n.d. https://history.nasa.gov/computers/Ch2-5.html.

———, ch. 2: "The Apollo Guidance Computer: Using the AGC." NASA History Office, n.d. https://history.nasa.gov/computers/Ch2-7.html.

———, ch. 4: "Developing Software for the Space Shuttle: Computers in the Space Shuttle Avionics System." NASA History Office, n.d. https://history.nasa.gov/computers/Ch4-5.html.

Corfield, Richard. "Apollo 13: Houston, we've had a problem." *Chemistry World,* March 2010, 56–61.

Dethloff, Henry C. *Suddenly Tomorrow Came: A History of the Johnson Space Center.* NASA History Office, 1993. https://www.jsc.nasa.gov/history/suddenly_tomorrow/suddenly.htm.

Duggins, Pat. *Final Countdown: NASA and the End of the Space Shuttle Program.* Gainesville: University Press of Florida, 2009.

Edwards, Owen. "Splendid Isolation: When the First Astronauts to Walk on the Moon Returned from Their July 1969 Lunar Expedition, They Were Confined to Quarters." *Smithsonian,* July 2004. http://www.smithsonianmag.com/science-nature/splendid-isolation-2482597/.

Erikson, Arthur. "The Changing Face of Engineering." *Electronics* 56, no. 11 (May 31, 1983): 125–33.

Hadfield, Chris. *An Astronaut's Guide to Life in Earth.* New York: Macmillan, 2013.

———. *You Are Here: Around the World in 92 Minutes.* Boston: Little, Brown, 2014.

Hansen, James R. *First Man: The Life of Neil A. Armstrong.* New York: Simon and Schuster, 2012.

Harwood, William. "Legendary Commander Tells Story of Shuttle's Close Call" (Astronaut Robert "Hoot" Gibson comments on the STS-27 near disaster and loss of Space Shuttle Columbia). Spaceflight Now, March 27, 2009. https://spaceflightnow.com/shuttle/sts119/090327sts27/.

Heath, David. *Living on the Edge: Proud to Serve, Fast-Paced Learning, Pleased to Share, Life-Long Travel.* Privately published, 2017.

"The History of Women in Computing." Florida Tech Online, 2017. https://www.floridatechonline.com/blog/information-technology/the-history-of-women-in-computing/.

Holland, Brynn. "Human Computers: The Women of NASA." History in the Headlines, December 13, 2016. http://www.history.com/news/human-computers-women-at-nasa.

"Index, Declassified Manned Orbiting Laboratory (MOL) Records." National Reconnaissance Office, n.d. http://www.nro.gov/foia/declass/MOL.html.

Keel, Bill. "Telescopic Tracking of the Apollo Lunar Missions." Bill Keel's Space Bits, University of Alabama Department of Physics and Astronomy, n.d. http://pages.astronomy.ua.edu/keel/space/apollo.html.

Kennedy, President John F. "I believe that this nation should commit itself to achieving the goal, before this decade is out, of landing a man on the moon and returning him safely to the earth." Special Message to the Congress on Urgent National Needs, May 25, 1961.

———. "In a very real sense, it will not be one man going to the moon . . . it will be an entire nation." Special Message to the Congress on Urgent National Needs, May 25, 1961.

———. "We choose to go to the Moon . . ." Address at Rice University on the Nation's Space Effort, September 12, 1962.

Kranz, Gene. *Failure Is Not an Option.* New York: Simon and Schuster, 2000.

Leveson, Nancy G. "Software and the Challenge of Flight Control." Ch. 7 in *Space Shuttle Legacy: How We Did It and What We Learned*, ed. Roger D. Launius, John Krige, and James I. Craig. Reston, Va.: American Institute of Aeronautics and Astronautics, 2013.

Lewis, Anna. "When Computer Programming Was 'Women's Work.'" *Washington Post,* August 26, 2011. https://www.washingtonpost.com/opinions/

when-computer-programming-was-womens-work/2011/08/24/gIQA-dixGgJ_story.html?utm_term=.d9fd019a5416.

Lewis, John F. *Mining the Sky: Untold Riches from the Asteroids, Comets, and Planets*. New York: Helix Books–Perseus Books Group, 1996.

Light, Michael, and Andrew Chaikin. *Full Moon*. New York: Alfred A. Knopf, 1999.

Lineback, Robert. "Code Check Speeds Launches." *Electronics* 56, no. 8 (April 21, 1983): 48–49.

Lovell, Jim, and Jeffrey Kluger. *Apollo 13*. Boston: Mariner Books–Houghton Mifflin Harcourt, 2000.

Michener, James A. *Tales of the South Pacific*. New York: Macmillan, 1947.

Moon Machines (mini-series, DVD set). Directed by Christopher Riley. Discovery Studio, Science Channel, 2008.

Mullane, Mike. *Riding Rockets: The Outrageous Tales of a Space Shuttle Astronaut*. New York: Scribner, 2006.

Mythbusters: "Are the Apollo moon landing photos fake?" Mythbusters, Discovery.com, n.d. http://www.discovery.com/tv-shows/mythbusters/mythbusters-database/apollo-moon-landing-pictures-fake/.

NASA Johnson Space Center: Oral History Project. History Portal, 1997–. https://www.jsc.nasa.gov/history/oral_histories/oral_histories.htm.

"NASA Spacecraft Images Offer Sharper Views of Apollo Landing Sites." NASA Johnson Space Center Mission Pages, Apollo Revisited, September 5, 2011. https://www.nasa.gov/mission_pages/LRO/news/apollo-sites.html.

NSU Receives NASA Space Shuttle On-Board Computer for Research. Press release, Nova Southeastern University, December 17, 2012. http://spaceref.com/news/viewpr.html?pid=39566.

Orwig, Jessica. "This Amazing 25-Year-Old Woman Helped Bring Apollo Astronauts Back from the Moon." *Business Insider*, December 9, 2014. http://www.businessinsider.com/poppy-northcutt-helped-apollo-astronauts-2014-12.

Platoff, Anne M. *Where No Flag Has Gone Before: Political and Technical Aspects of Placing a Flag on the Moon*. NASA Contractor Report 188251, 1993. https://www.jsc.nasa.gov/history/flag/flag.htm.

Reichhardt, Tony, ed. *Space Shuttle–the First 20 Years*. New York: DK Publishing, 2002.

Report of Columbia Accident Investigation Board, vol. 1. Washington, D.C.: NASA, August 26, 2003. https://www.nasa.gov/columbia/home/CAIB_Vol1.html.

Science News, https://www.sciencenews.org/.

Scott, David Meerman, and Richard Jurek. *Marketing the Moon: The Selling of the Apollo Lunar Program*. Boston: MIT Press, 2014.

Shepard, Alan B. Jr. Interview by Roy Neal, Pebble Beach, Florida, February 20, 1998. Edited Oral History Transcript, NASA Johnson Space Center Oral History Project. https://www.jsc.nasa.gov/history/oral_histories/Shepard-AB/shepardab.htm.

Siddiqi, Asif A. *The Soviet Space Race with Apollo*. Gainesville: University Press of Florida, 2003.

Space Shuttle: Orbital Maneuvering Systems. NASA Human Spaceflight, April 2002. https://spaceflight.nasa.gov/shuttle/reference/shutref/orbiter/oms/.

Space Shuttle Propulsion Systems. NASA Document FS-2005–04–028-MSFC, April 2005. https://www.nasa.gov/sites/default/files/174533main_shuttle_propulsion.pdf.

Spector, Alfred, and David Gifford. "A Case Study: The Space Shuttle Primary Computer System." *Communications of the ACM* 27, no. 9 (September 1984): 874–900.

"The Texas Water Quality Board Took Steps . . . for the Clear Lake Watershed." *Lubbock Avalanche-Journal*, June 27, 1969, 3. https://www.newspapers.com/newspage/6129238/.

Tregaskis, Richard. *X-15 Diary: The Story of America's First Space Ship*. Lincoln: Bison Books–University of Nebraska Press, 2004.

Van Dusen, Matthew, ed., Txchnologist.com. "Will Anyone Recover Apollo 13's Plutonium?" *Space Safety*, June 2, 2014. http://www.spacesafetymagazine.com/aerospace-engineering/nuclear-propulsion/will-anyone-recover-apollo-13s-plutonium/.

Wikipedia. Third-Party Evidence for Apollo Moon Landings. https://en.wikipedia.org/wiki/Third-party_evidence_for_Apollo_Moon_landings#Apollo_13.

INDEX

Page numbers in *italics* refer to photos and illustrations.

Jack Clemons is a professional writer and consultant as well as a speaker and presenter on NASA's space programs. Earlier in his career he was an engineering team leader on NASA's Apollo Program and senior engineering software manager on the Space Shuttle Program in Houston, Texas. He has also held the role of Senior Vice President of Engineering for Lockheed Martin. He has bachelor's and master's degrees in aerospace engineering from the University of Florida. His works of fiction have earned him an Established Artist Fellowship Grant for Literary Fiction by the Delaware Division of the Arts and membership in the Science Fiction and Fantasy Writers of America.